Societies under Construction

Daniel J. Sage · Chloé Vitry
Editors

Societies under Construction

Geographies, Sociologies and Histories of Building

Editors
Daniel J. Sage
School of Business and Economics
Loughborough University
Loughborough, UK

Chloé Vitry
School of Business
University of Leicester
Leicester, UK

ISBN 978-3-319-73995-3   ISBN 978-3-319-73996-0   (eBook)
https://doi.org/10.1007/978-3-319-73996-0

Library of Congress Control Number: 2018939714

Cover credit: © Naufal MQ/Getty Images

Printed on acid-free paper

This Palgrave Macmillan imprint is published by the registered company Springer International Publishing AG part of Springer Nature
The registered company address is: Gewerbestrasse 11, 6330 Cham, Switzerland

Acknowledgements

This book was inspired by a workshop on 'Building Work: Histories, Geographies and Sociologies' held on the 10th October 2014 at Loughborough University. This workshop was made possible with the kind support of the Centre for Professional Work and Society in the School of Business and Economics, and the School of Civil and Building Engineering. While not all of the contributors to this book attended that workshop, the lively and fruitful discussions at the workshop around the diversity and future of social studies of building construction persuaded us to develop more outlets for social science and humanities debates about building construction. We would also like to thank Peter Ackers, Andrew Dainty and Christine Coupland who, as well as being influential in supporting the workshop, encouraged us more widely as to the unique value of edited interdisciplinary collections. And, finally, we must also acknowledge the considerable efforts of all the contributors to this volume, and the guidance of Holly Finch and Joanna O'Neill at Palgrave.

Contents

Editors and Contributors

About the Editors

Dr. Daniel J. Sage is Senior Lecturer in Organizational Behaviour in the School of Business and Economics at Loughborough University, United Kingdom. His research focuses upon the development of theories of materialities, geographies and power within organizational life, particularly in relation to the built environment. He has produced over 30 publications across a wide variety of social science disciplines and international journals.

Dr. Chloé Vitry is a Teaching Fellow in the School of Business at University of Leicester, with a Ph.D. in Organization Studies. She previously worked as a Research Associate at Loughborough University's School of Business and Economics, researching the travel of innovations in the construction industry. She is interested in bodies and sociomateriality, affect and alternative forms of organizing. She is a member of VIDA, a Critical Management Studies association.

Contributors

Paul Chan is Senior Lecturer in Project Management in the School of Mechanical, Aerospace and Civil Engineering at the University of Manchester, United Kingdom. Paul studies human relations in engineering and construction project contexts, focussing particularly on how people cope with social, organizational and technological change. He is especially interested in the everyday mundane practices and processes of ongoing change in organizational life. He is one of the Editors of *Construction Management and Economics*. He co-authored *Constructing Futures* (Wiley-Blackwell, 2010), and over 80 peer-reviewed journal and conference articles.

Andrew Dainty is Professor of the Sociology of Construction and is based in the School of Architecture, Civil and Building Engineering, Loughborough University, UK. For the past 22 years he has researched the sociologies of construction practice, focusing on the social rules and processes that affect people working as members of project teams. He has also led work mobilising critical perspectives on projects and the management people within the construction sector. He has published over 350 papers and is co-author/editor of eight books and research monographs.

Dr. Monika Grubbauer is Professor in History and Theory of the City at HafenCity University, Hamburg, Germany. She is a trained architect with a Ph.D. in social and economic sciences. She is interested in the interdependencies between economic, social and material processes of urban change. She has published widely on urban transformation and socio-economic restructuring in different geographical contexts, with a particular focus on the role of architecture, planning and construction. Her recent work explores processes of internationalization, marketization and financialization in construction and retailing in Mexico and the consequences for informal settlements and self-organized building practices.

Dr. Lise Justesen is Associate Professor, in the Department of Organization at Copenhagen Business School, Denmark. Her research interests encompass organizational value, critical accounting studies, technologies, materialities, and sustainability. She has published on these subjects in numerous journals, including: *Organization, Construction Management and Economics, Accounting Auditing and Accountability Journal* and *Accounting, Organizations and Society*.

Dr. Martin Löwstedt is a Senior Lecturer at the Department of Architecture and Civil Engineering, Chalmers University of Technology, Sweden. His research areas concern aspects of professional identity, leadership, and strategy practices, focusing on the Swedish construction industry. His major publications in English are: "Social identity in construction: enactments and outcomes" (*Construction Management and Economics*, 2014) and "Taking off my glasses in order to see: exploring practice on a building site using self-reflexive ethnography" (*Construction Management and Economics*, 2015).

Jan Mouritsen is Professor in the Department of Operations Management at Copenhagen Business School, Denmark. His research interests encompass critical perspectives of accounting, practices and infrastructures of valuation, materiality, organizational learning, intellectual capital, and technology and innovation. He has produced over 100 conference papers and publications on these and other topics.

Dr. Jacob Paskins teaches architectural history and theory at University College London, United Kingdom. His research focuses on architecture and urban development in post-war France and Britain. Jacob is author of *Paris Under Construction: Building Sites and Urban Transformation* (Routledge, 2016), which he completed during a research fellowship at Girton College, University of Cambridge. He is currently working on a project about hoverports and high-speed marine travel during the second half of the twentieth century.

Dr. Hiral Patel's recently completed Ph.D. research, in the School of the Built Environment at the University of Reading (United Kingdom), theorizes the practices of adapting buildings. Drawing from Science and Technology Studies, her approach to studying buildings over time combines a range of ethnographic and historical methods. She is particularly interested in developing the practice of curating exhibitions as part of the analytic research process. Her current research around the archive of the DEGW architectural practice furthers this interest and explores the linkages between organizational practices and the built environment to help understand the changing nature of 'work'.

Dr. Ani Raiden is a Senior Lecturer at Nottingham Business School, Nottingham Trent University, United Kingdom. Her research on well-being, work-life balance, quality of working life, strategic and international HRM, and construction and project management has been published at major international and national conferences, and leading journals in the field. Ani is a Chartered Member of the Chartered Institute of Personnel and Development (CIPD), and the Immediate Past Chair of the Association of Researchers in Construction Management (ARCOM).

Dr. Christine Räisänen is Professor Emerita of Organization and Communication at the Department of Architecture and Civil Engineering, Chalmers University of Technology, Gothenburg, Sweden. Her research interests are organization studies, leadership, project management, cross-cultural and cross-professional interaction and communication, work-life balance and gender studies. Her background is in literary theory and applied linguistics. She has published her work in a number of journals, spanning a variety of fields, has convened thematic tracks at international conferences such as APROS and EGOS.

Rikard Sandberg is a Ph.D. candidate in Construction Management at Chalmers University of Technology, Gothenburg, Sweden. He did his Master's degree in the field of work sociology. His Ph.D. project is interdisciplinary, and concerns the everyday work of middle managers in the construction industry and how it implicates their work and life situations. His research interests concern critical perspectives on work-life balance, well-being, identities and autonomy/control.

Dr. Kjell Tryggestad is full professor at Inland Norway University of Applied Sciences, Department of Business Administration, Norway and associate professor at Copenhagen Business School, Department of Organization, Denmark. Research interests spans: project- and construction management, organization theory/actor-network theory, management accounting and innovation. His approach is processual with case studies of; strategic- and organizational change, design of new products, services, work practices, technologies, and more specifically, conflicting matters of concern related to sustainable development and the built environment. He has published widely on these topics in books and journals within construction management, accounting, organization studies, sociology and project management.

Dr. Dylan Tutt is Lecturer in Workforce Studies in Construction in the School of the Built Environment at University of Reading, United Kingdom. His qualitative research focuses on social interaction, communication and technology in complex organizational environments, although he maintains a wider interest in quite diverse areas of cultural sociology, stemming from his background in sociology and visual culture. Dylan specializes in ethnographic and video-based studies of communication and (work) practice in diverse settings. Most recently, Dylan has been pursuing his strong research interest in migrant workers, construction employment patterns, and the practice and management of occupational health and safety.

List of Figures

1

Introduction: Societies under Construction

Daniel J. Sage and Chloé Vitry

To the open possibility
of steel against sky
we weld, bolt and strap

wide staircases of marble, arched
skylights, commanding views

serviced by windowless corridors
where ceilings hang low, as though
the ones who will push carts and carry trays
are unusually small or,
prefer to scurry like mice
in closed dark spaces

or, as though
extra headroom might give them
                    ideas.

D. J. Sage (✉)
School of Business and Economics, Loughborough University,
Loughborough, UK
e-mail: d.j.sage@lboro.ac.uk

© The Author(s) 2018
D. J. Sage and C. Vitry (eds.), *Societies Under Construction*,
https://doi.org/10.1007/978-3-319-73996-0_1

Following the Blueprints, used with permission © 2006 Susan Eisenberg (Eisenberg 1998a, p. 3).

When walking through most large urban landscapes, the ubiquity of building construction is highly visible, from skylines intersected with tower cranes to streets darted by trucks carrying building materials and waste, hoardings diverting pedestrians, buildings surrounded by scaffolds of stainless steel or bamboo, and the noise of shouted instructions and pneumatic drills. Construction is neither an inconspicuous, unusual, nor insignificant presence in urbanizing societies. Yet in many ways its ubiquity presages its elision: building sites readily become a taken-for-granted part of the urban landscape, a source of curiosity for some perhaps, of annoyance and disruption for others and simply unnoticed by many more. As Glasser (2008) recognizes: 'For pedestrians and residents, they mean noise, dirt and dust – usually over a period of months. We often pass them holding our breath with our eyes averted, hoping that this apparition will soon be exorcised' (p. 16). While it is surely a truism that societies are made possible by building (Clarke 1992; Gieryn 2002), society is also not readily given to be captivated by such processes. That is, the transformative social role of building work is partly premised on eliding its capacity to bring about such profound changes. Architectural plans and usable buildings are fully intended to inspire, distract and transform a complex milieu of meanings, feelings and behaviours in users, cultural critics, artists, thinkers, publics and so on (Beauregard 2015; Lees 2001; Whyte 2006); yet, whether for good or ill, the transformative, often unruly role of what is usually termed 'construction', 'building' or 'building construction',[1] appears intended to proceed with little social attention and cultural register.

C. Vitry
School of Business, University of Leicester, Leicester, UK
cav5@le.ac.uk

[1]To avoid excessive repetition across this introduction, we interchangeably use the terms 'building construction' as well as 'building' (as a verb), 'building work' and 'construction' to refer not only to those rather singular activities of building implementation, as set out in industrial classifications (see ISIC 2017b; UKSIC 2017), but also reflect upon how those activities are socially organized in multiple, changing, ways as an object of knowledge (Foucault 2004).

Underpinning the everyday elision of processes of construction is a set of corresponding intellectual characterizations. For centuries, building work has been readily been mobilized in Western intellectual thought to define means/ends thinking, or 'instrumental rationality', wherein thought is reduced to the 'rational consideration of alternative means to the end, of the relations of the end to the secondary consequences, and finally of the relative importance of different possible ends' (Weber 1978, p. 26). Plato, in *The Stateman* (circa 360 BCE), offers the earliest recorded example of this proclivity when he discusses building work to define 'directive' knowledge contrasted with the nobler 'calculative' knowledge within fields such as philosophy, rhetoric and mathematics, concerned with evaluative verdicts on such work (Plato 1961, p. 127). For Plato, the knowledge of the 'master builder' is limited to giving 'the appropriate directions to each of the workmen and see that they complete the work assigned' (Plato 1961, p. 126). The legacy of the Platonic treatment of building work to define instrumental knowledge appears profound. It readily inflects the cleavage of architectural thinking from 'more physical' building work (Pont 2005), and by extension the classed division of labour between those professions involved in undertaking building design and the building trades (Clarke et al. 2013; Thiel 2007, 2010, 2013). Defined in terms of the instrumental application of 'higher' knowledge, and serving mostly managerial, financial and technical ends, building construction is readily debased as a place to develop novel understandings of human societies, histories and geographies.

It is set against the shadow of such historical framings that contemporary ways of knowing building construction have emerged. And unsurprisingly, knowledge which promises to be easily instrumentalized to achieve pre-given managerial, financial and technical ends, as that from positivistic-predictive academic fields, such as engineering and management science, and commensurate social science fields, such as economics and psychology, has served as the primary basis for the emergence of distinct academic sub-disciplines, departments and research centres, such as construction engineering and management (Dainty 2008; Seymour and Rooke 1995). Over recent decades, the narrowness of such disciplinary framings has not passed without some critical redress, particularly in terms of the limitations it presents in understanding complex social relations (Seymour et al. 1997). Since the turn of millennium, Anglophone construction management research (CMR) has evidenced a growing interest in perspectives from disciplines across the social sciences and humanities (Harty and Leiringer 2017; Schweber 2015). However, such wider disciplinary engagements often themselves remain framed in decidedly instrumentalist terms: theories from social science and humanities disciplines, such as geography, sociology and history, are commonly applied to address challenges stemming from the construction industry and its stakeholders, such as technological innovation (Harty 2008), employee discrimination (Smith 2013) and occupational health and safety (Sherratt 2015). Reciprocally, relative to other aspects of the built environment scholarship, such as that related to architectural design or building use, building construction remains a comparably seldom visited object of analysis for academics based within social science and humanities departments and research groups. Equally, social science and humanities studies that do theorize with building construction (e.g. Hayes 2002; Kraftl et al. 2013; Thiel 2010) often remain overlooked within the fields of construction engineering and management. These disciplinary disconnections have not entirely passed without notice, as witness recent calls by prominent construction management researchers to contribute to wider debates in the social sciences and humanities (e.g. Bresnen 2017; Harty and Leiringer 2017; Leiringer and Dainty 2017; Styhre 2017). However, by and large, such calls serve mostly to evidence the unharnessed potential of building construction to inspire new ways of understanding, and perhaps radically changing, societies.

The purpose of this collection is to offer a sustained acknowledgement, understanding and challenge to the marginalization of construction as a place to develop novel understanding of society. In so doing, we seek to unpack and disavow everyday and intellectual characterizations that demarcate construction work as socially and culturally unremarkable. This book brings together authors that are concerned with theorizing human societies with construction across three disciplinary fields: sociology, human geography and history. These disciplines have been chosen as scholars working within them have long been concerned with the intersection of society and the built environment and are also typically more inclined to suspend the instrumentalist framing of knowledge that typifies engineering and positivistic social and management sciences (Fournier and Grey 2000; Lyotard 2002). By setting up the rationale for this volume in this way we also recognize the potential for exacerbating rifts between scholarly communities. For this reason, our collection includes scholars working from across the social sciences and humanities, and within construction management and engineering departments. Our aim here being to both encourage scholars within construction management and engineering research groups, departments and schools to contribute to the wider development of theories in social science and humanities disciplines, such as sociology, human geography and history, and to encourage scholars within the latter disciplines to theorize with construction. We also anticipate such contributions to theory development with construction in the social sciences and humanities *may* indirectly lead to substantive reframings of more managerial research concerns and agendas within construction management and engineering communities. After all, as will be discussed shortly, the history of intellectual framings and reframings of construction as a particular object of knowledge, from Plato onwards, have influenced many of the most pressing concerns facing the construction industry today.

Our introduction to this collection develops across four sections. First, we elaborate how construction has been historically formed as a largely instrumentalist object of knowledge across relatively enduring discourses spanning Western intellectual thought. The purpose of this discussion is to uncover how this framing endures to circumscribe construction as an object of knowledge in instrumental terms serving technical, managerial, physical and financial ends. Then, second, we explore the 'social turn' in Anglophone CMR as a potentially welcome response to this situation and the limitations it presents in understanding social relations within and around building construction. Emerging across the last two decades or so, authors working within this 'social turn' have argued for more sustained engagements between construction scholarship and social science and humanities disciplines. However, while seemingly disrupting the disavowal of construction as an object of scholarly interest for the social sciences and humanities, the predominantly applied nature of such engagements, where social theories are mostly applied not developed, limits its potential to challenge long-standing framings of building construction as socially unremarkable. Third, we turn to studies of the built environment in human geography, sociology and history, to explore the increasing, if still marginal and fragmented, receptivity to considering the places, people and politics of building construction within those disciplines. Finally, we reflect back on the specific rationale for this interdisciplinary volume and elaborate on the range of ways construction might be relocated as a richer object of knowledge for human societies by introducing the seven contributory chapters that constitute this collection.

The Formation of Construction as an Object of Knowledge

Building construction is typically defined in terms of a set of largely managerial, technical and financial, and physical, activities involved in generating economic value related to the assembly of new buildings and structures. The International Standards on International Classification (ISIC), produced by the United Nations (ISIC 2017a), is exemplary in this regard and also underpins many national and supranational classifications, such as the UK's Standard Industrial Classification (UKSIC 2007). ISIC defines construction in 'Section F' as 'new work, repair, additions and alterations, the erection of prefabricated buildings or structures on the site and also construction of a temporary nature' (ISIC 2017b). This definition encompasses five sub-sectors: 'site preparation' (demolition of old structures, site clearance, earth moving), 'building of complete constructions or parts thereof', civil engineering' (i.e. construction of all buildings and structures and specialized structure skills such as bricklaying and steel erecting), 'building installation' (e.g. plumbing, wiring and electrical work), 'building completion' (e.g. plastering, painting and decoration) and 'renting of construction or demolition equipment with operator' (renting plant equipment with operator). The UKSIC, adapted from ISIC, renders the definitional emphasis on financial, managerial and technical activities more explicit still, specifying construction as the 'financial, technical and physical means to realise the building projects for later sale' (ONS 2017). Significantly, under the ISIC (and UKSIC) both building engineering and architecture are classified separately.

What is striking across these industrial classifications is that they repeat a decidedly Platonic view of building construction as a process of instrumental (or 'directive'), technical, financial and managerial effort, separate from the seemingly more 'calculative' domains of building design, architecture and engineering, and its underpinning philosophical, artistic and mathematical knowledge. As such, we contend that these industrial classifications are far from objective and passive categories, rather they actively operate to form 'construction' as an instrumental object to 'better' apply technical, managerial and financial knowledge to serve technical, managerial and financial ends. As Foucault (2004) explains, objects of knowledge, such as building construction, never pre-exist the discursive relations—whether industrial classifications or philosophical treatises—through which they are known, rather they are formed through these relations. Such discourses correspond to historically accumulated relations of delimitation and specification, involving specific techniques, institutional authorities and social norms, where objects of knowledge are differentiated allowing them to become 'manifest, nameable, and discernible' to certain forms of knowledge (Foucault 2004, p. 46). This Foucauldian view of objects of knowledge as relationally and historically constituted offers a useful way of understanding how construction has emerged to enable particular forms of knowing the world, and with specific social consequences. It is beyond the scope of this introduction to offer a comprehensive archaeology of the formation of construction as an object of knowledge; instead, we will specify some significant historical reference points to get a better grip on how construction research has been devalued within the social sciences and humanities. Much of what follows in this section concerns one of the most institutionalized sets of discursive relations around construction: the rise of definitional distinctions between 'construction' and 'architecture'.

While the terms 'architecture' and 'construction' are typically understood separately today (as in ISIC 2017), they both share a common etymological origin in the ancient Greek word *architecton*—translatable as 'master builder'. This is why in the passage cited at the head of this introduction, Plato seemingly avoids any distinction between design and building, reducing all building to the coordination of manual effort, the instrumental implementation of ('calculative') knowledge, whether from philosophy, rhetoric or mathematics to achieve a given building task (Plato 1961). Plato's thesis presents the obvious challenge that building often involves the development of new theories in such 'noble' disciplines. After all, the increasing urbanization of humanity is hardly incidental to shifts in the history of human thought (Meagher 2008) and all making, all craft, surely involves new thought (Ingold 2013; Sennett 2008). However, instead of presaging an intellectual aggrandization of building construction the intellectual response to Plato's narrowly instrumental view of building seemingly ushered the reverse: in order to redress the Platonic circumscription of building as *architecton*, this object of knowledge was progressively bifurcated into two elements: 'architecture' and 'construction', each with a new set of professional roles, interests and knowledges. The earliest semblance of this split can be traced to the Roman architect Vitruvius and his depiction of the architect in *De Architectura* (written circa 15 BCE):

> Let him be educated, skilful with the pencil, instructed in geometry, know much history, have followed the philosophers with attention, understand music, have some knowledge of medicine, know the opinions of the jurists, and be aquatinted with astronomy and the theory of the heavens. (De Architectura, Book I:3, Vitruvius 1960, p. 5)

At first glance, Vitruvius seemingly challenges Plato by defining architecture as the fusion of practical and theoretical knowledge. Yet crucially, his understanding of practical knowledge is expressed not in terms of the physical effort of building or even labour coordination, as with Plato, but in how the sciences (physics, medicine, geometry and optics) can inform building design, alongside philosophical, historical, legal and artistic knowledge. *De Architectura* does contain detail on the techniques involved in the act of construction itself—such as the laying of appropriate foundations for temples—and indeed even contains guidance regarding the delivery of buildings to agreed budgets. However, as Pont (2005) explains, the notion of architectural practice (or 'Fabrica') as used by 'Vitruvius, does not denote making, practical building or the art of construction; and, most emphatically, that it does not refer directly to any kind of manual art' (p. 77). Rather:

> Fabrica, for Vitruvius, is the practice of architectural contemplation. By extension it comes to embrace both the process and the product of architectural contemplation – the practical 'know how' of surveying, site-analysis, design, planning, calculation and management, rather than the muscular and intuitive grasp of actual building that few architects experience today. (Pont 2005, p. 78)

In other words, in *De Architectura*—the first known treatise on architecture—the prosaic realities of labour coordination and craftwork (Ingold 2013; Sennett 2008) are deemed unworthy of codification, thus located as not only distinct, but inferior, to the professedly more reflective, rarefied, activity of building design and engineering.

As has often been noted (Ingold 2013; Yarrow and Jones 2014), such architectural self-aggrandizing was notably absent during the medieval period, where the design work involved in the construction of Gothic cathedrals can be viewed as tightly connected to manual effort. However, it re-emerged prominently as an intellectual discourse, and accompanying division of labour, during the renaissance of classical thought. This was particularly evident in Italian city states, especially Florence in the wake of the construction of the Duomo di Firenze at the Basilica di Santa Maria del Fiore—the dome of Florence at St. Mary of the Flower Cathedral—completed in 1436. The building of the Duomo, led by Fillipo Brunelleschi, remains the largest self-supporting dome in the world constructed entirely of brick, timber and mortar. It was made possible by Brunelleschi's knowledge of Roman building engineering techniques as detailed in *De Architectura* (a text republished in Florence in 1414) and by studying the Roman Pantheon dome in Rome (King 2000). Temporary scaffolding and hoists, powered by oxen, were used to move large amounts of masonry, while precise geometric measurements were taken to design the double-shell ring and rib support for the dome and herringbone pattern for the brickwork. These techniques made it possible to construct the dome without the need for permanent structural support and buttresses, as used in Gothic cathedrals, as had been assumed to be mandatory to any attempt to construct the dome (Battisti 1981; King 2000).

The assemblage of the Duomo outwardly fits with the Vitruvian ideal of the architect—the embodied fusion of precise mathematical knowledge and an artistic appreciation of pure (Platonic) geometric forms. The inscription on Brunelleschi's tomb in the Basilica di Santa Maria del Fiore certainly pays homage to this lofty ideal:

> One whose counsel was sought everywhere, a man of exceptional ingenuity in solving difficult problems or, indeed, in recognizing their existence, and the inventor of machines of enormous innovative importance for the future – machines which made it possible to substitute intelligence for brute force, and meticulous execution for physical effort and improvisation. (translated by Battisti 1981, p. 9)

However, Brunelleschi was also present on site during the construction of the dome, helping workers reapply Roman construction techniques (King 2000), including the painstaking creation of construction machines that safely and securely lifted the 25,000 tons of materials for the construction of the Duomo (Battisti 1981, pp. 9, 114). Brunelleschi organized day-to-day work on site and engaged in the messy and prosaic politics and events that ensued: he hid aspects of his design to maintain control over the project; he sacked workers to re-hire them at lower rates of pay; and he also left the project at one stage to demonstrate the incompetence of his partner, Ghiberti (Battisti 1981; King 2000). In short, Brunelleschi was far from simply a distant aesthete; yet his work, and that of his renaissance contemporaries, offered a seductive touchstone for inspiring a self-aggrandization of architecture as a noble intellectualized, socially meaningful, profession, capable of developing new social knowledge, by distinguishing it from the manual effort of building as a relatively unthinking, instrumental, pursuit. This sentiment is perhaps most vivid in the thinking of his younger friend, and fellow architect, Leon Alberti. In the prologue of his reimagining of Vitruvius's in the architectural treatise *On the Art of Building in Ten Books* (1452), Alberti elaborates:

... I should explain exactly whom I mean by an architect; for it is no carpenter that I would have you compare to the greatest exponents of other disciplines: the carpenter is but an instrument in the hands of the architect. Him I consider the architect, who by sure and wonderful reason and method, knows both how to devise through his own mind and energy, and to realize by construction, whatever can be most beautifully fitted out for the noble needs of man, by the movement of weights and the joining and massing of bodies. To do this he must have an understanding and knowledge of all the highest and most noble disciplines. This then is the architect. (Alberti 1988, p. 3)

Alberti, as with Plato, here mobilizes the building labour process to draw a distinction between physical and mental labours, the latter being the mere physical instrument for the thought of the former. This intellectual debasement of construction site work is especially salient: by also being the first to codify the linear perspective, and its purportedly more 'truthful' way of seeing the material world, Alberti's thought had a profound influence on the development of rationality, objectivity, scientific truth, individualism and capitalism, during the Enlightenment (Gregory 1994, pp. 390–395; Harvey 1989, pp. 244–245), including Descartes famous affirmation of priority of the thinking human *cogito* over the inert material world of bodies, things and animals (Purdy 2011, pp. 87–89). After the Enlightenment, the use of the construction site as an unthinking 'Other' to architecture to mark out professedly nobler, or enlightened, interests, knowledges and professions became increasingly commonplace—consider, for example, the French architectural theorist, Marc-Antoine Laugier in his *An Essay on Architecture* (1753):

When one speaks of the art of building, the chaotic mess of clumsy debris, immense piles of shapeless material, a dreadful noise of hammers, perilous scaffolding, fearful grinding of machines and an army of dirty and mud-covered workmen—all this come to the mind of ordinary people, the unpleasant outer cover of an art whose intriguing mysteries, noticed by few people, excite the administration of all those who penetrate them. There they discover inventions of a boldness [the architect] that proclaims a great and fertile genius proportions of stringency that indicates severe and systematic precision, and ornaments of an elegance that tells of a delicate and exquisite feeling. (Laugier 1753/1977, p. 7)

At the heart of such definitional proclamations appears to be what the German historian Daniel Purdy proposes a feeling that 'architectural treatises from Vitruvius on have struggled against ... [the] ... anxiety that architects are no more than elevated masons' (Purdy 2011, p. 37). This anxiety, itself driven by Plato's intellectual debasement of all building, has seemingly vexed architectural thinkers for centuries. However, the resultant solution: the intellectual denigration of construction work is seldom acknowledged let alone critically interrogated (for a brief exception, see Ingold 2013, pp. 49–59). Whether overlooked completely as with Vitruvius, depicted as prosaically instrumental, and socially unremarkable, as in Alberti and Plato, or designated a place of sheer chaos and disgust as in Laugier, the places, peoples and politics of construction, as objects of knowledge, have purposefully, variously and repeatedly been shorn of wider significance in understanding human societies.

It is against this historical backdrop that we can start to better understand, and thus challenge, the separation of architecture and construction in industrial classifications. Two objects of knowledge—architecture and construction—have been formed over centuries to instil a historically significant set of interlinked binary oppositions: calculate/directive knowledge, designing/making, signification/signified, theory/practice, thought/matter, subject/object, mind/body and even human/non-human.[2] These intellectual separations have had profound implications for how buildings are built and how societies are reproduced through that process. In particular, they engender a strong separation between the professional education and careers of architects, and those involved in 'implementing' those plans, whether construction managers, engineers or those in the building trades. In the UK, for example, architecture requires five years of study across a two-stage university degree and two years of practical experience to be

---

[2]The non-human/human binary can be derived from the architecture/construction opposition in two ways. First, in some accounts building labour is framed as non-human, as in Alberti's (1988) treatment of the building trades as a passive 'tool' of the architect. This dehumanization resembles recent sociologies (e.g. Thiel 2013) where those working in building trades are described by themselves and colleagues as 'tools'. Second, the Cartesian 'ontological separation of mind and matter reenforces a binary distinction between the rational thoughts of a subject and the predictable behaviour of material bodies' (Roberts 2012, p. 2514).

registered to practice by the Architectural Registration Board; these programmes encompass a fairly broad mix of disciplinary influences from across the physical sciences, as well as the social sciences and humanities. In contrast, construction trade roles, such as bricklaying, emphasize practical experience. Indeed, although formal qualifications in bricklaying are available in the UK, these focus narrowly on 'embodied labour' (physical bricklaying techniques) over wider managerial, environmental and social knowledge (Clarke et al. 2013). What is more, construction industry site and project managers are typically promoted from the building trades and/or gain management and engineering focussed construction management degrees—few possess architectural degrees. Heeding the Platonic legacy in architectural thinking, UK construction engineering and management degrees are themselves strongly steeped in (technical and managerial) instrumental rationality and, relative to architectural degrees, typically lack engagement with theories around the significance of building from the social sciences and humanities. Given the classed and gendered coding of knowledge, education and careers (Bourdieu 1984; Willis 1977), these differences in professional experiences inevitably imbue social differentiations between a more broadly educated, likely more middle-class, comparatively more female,[3] architectural profession and a narrowly, technically and/or managerially qualified, likely more working-class, comparatively more male, construction workforce. Such separations in training have been further institutionalized in the organization of building work as architects often play a minimal, increasingly remote role in the on-site delivery of buildings, despite the ever-present need to adapt designs around emergent events, conditions, uses and users (Lorne 2017; Moran et al. 2016; Sage 2013). Equally, construction contractors are seldom involved in early design decisions, and discussions about wider urban planning and development, despite long-established concerns around the fragmentation of the building process, not least between 'design' and 'build' stages (Lahdenpera 2012).

---

[3]While architecture is certainly not free from gendered discrimination (Powell and Sang 2015), it is often less male dominated; for example, in the UK 25% of architects are female, with significant recent increases (Waite 2017), while only 1% of workers on site are women (UCATT 2017).

Many of these industrial challenges have been the subject of extensive inquiry in academia and beyond, but what is less readily acknowledged is that they all can be located within centuries-old emergence of construction as a particular (i.e. instrumental) object of knowledge, distinct from substantive reflections on knowledge, not least those related to human history, geographies and societies. The purpose of this edited collection is to understand, and challenge, this situation, but in order to do so we will first consider how a growing number of construction scholars have reflected upon, and challenged, the disconnection between construction and social science and humanities thinking that has for so long framed how building construction is to be understood as an object of knowledge. While in many ways commensurate with the purposes of this book, this body of work also evidences limitations that will be addressed.

The 'Social Turn' in Construction Management

In the *Postmodern Condition: A Report on Knowledge* Lyotard (2002) explains how the production of knowledge in technoscientific fields, such as construction, has become gripped by a 'generalized spirit of performativity' driven by an 'equation between wealth, efficiency and truth' (p. 45):

> A technical apparatus requires an investment; but since it optimizes the efficiency of the task to which it is applied, it also optimizes the surplus-value derived from this improved performance. All that is needed is for the surplus-value to be realized, in other words, for the product of the task to be sold. And the system can be sealed in the following way: a portion of the sale is recycled into a research fund dedicated for further performance improvement. (Lyotard 2002, p. 45)

Lyotard (2002) goes on to discuss how under capitalism an ecosystem of research in engineering and science has developed in which technological corporations fund their own research departments where demands for immediate performativity (increased surplus value—i.e. profit, productivity gains) are high, while also 'creating grant program subsidies to university departments, research laboratories, and independent research groups with no expectation of an immediate return on the results of the work' (p. 45). As Lyotard (2002) explains, these latter programmes are tolerated by corporations and the State as it is expected that some financial losses may be required in order to increase the probability of higher gains in profitability, which is assumed to underpin social progress. While Lyotard (2002) only considers technoscience, Fournier and Grey (2000) extend this 'performativity thesis' to positivistic management science which, they contend, has similarly become enthralled by an 'intent to develop and celebrate knowledge which contributes to the production of maximum output for minimum input [and] inscribing knowledge within means–ends calculation' (Fournier and Grey 2000, p. 17). Of course, such sweeping characterizations of engineering and management research as ideologically capitalistic, and epistemologically rationalist-positivist, are inevitably far from universal, as the emergence of Critical Management Studies demonstrates (Alvesson and Wilmott 1993; Alvesson et al. 2009). Nevertheless, within such fields it is hardly difficult to evidence the clamour for means-end thinking, wherein engineering and management knowledge is developed to improve the efficiency and effectiveness through which a problem, given in industry, and mostly defined by senior managers, can be solved (Feenberg 1999; Klikauer 2013; Parker 2002).

Given that construction has been discursively formed for centuries to define instrumental knowledge, it is not surprising that these more instrumentalist, problem-focused, disciplines have been central to the development of construction-focused academic disciplines—witness, for example, the institutionalization of construction engineering and management departments, journals and conferences (Harty and Leiringer 2017, p. 393). Whether evidenced in management studies of Critical Success Factors for private–public infrastructure projects (Zhang 2005), or the consequence of different construction strategies (Cheah et al. 2007), engineering studies of the viability of non-destructive testing (McCann and Forde 2001) or the use of stainless steel (Baddo 2008), construction research is frequently, if not predominantly, premised on improving the means by which given industry problems can be solved (cf. Schweber 2015). And, in keeping with Lyotard (2002), these ends are frequently defined in avowedly capitalistic and managerialist terms, for example, improvements in profitability, fiscal value, return on investment, competitiveness, cost-efficiency and productivity—all criteria that closely mirror the performance indicators pursued by managers in construction firms (Sage et al. 2013). Notwithstanding this prevailing intellectual emphasis, over recent decades, particularly in Anglophone CMR, there has been an increasing recognition of the limitations of developing knowledge about construction from such an overly narrow disciplinary base. This work has resulted in efforts to engage with social science disciplines, particularly sociology, that offer possibilities to critically reflect upon, perhaps even challenge the framing of building construction as a purely instrumental, or applied, object of knowledge.

One of the earliest, and most influential, events in this 'social turn' was the publication of a paper in *Construction Management & Economics* on the culture of construction research by Seymour and Rooke (1995) and a series of ensuing debates (e.g. Raftery et al. 1997; Harriss 1998; Seymour et al. 1997; Rooke et al. 1997; Runeson 1997; Seymour et al. 1998). At the heart of Seymour and Rooke's (1995) argument is a lament that CMR has inherited from its engineering origins a mostly rationalistic (quantitative, positivistic) world view which presumes that social research can deliver objective, nomothetic, knowledge of the causes of the construction process, independent of the meanings given by its participants. The purpose of this dominant approach is 'to construct a blueprint of how the [construction] process can be made to work most efficiently ... Offered as the pure expression of instrumental rationality it brooks no objection' (Seymour and Rooke 1995, p. 513). Setting out their critique, mostly with reference to ethnomethodology, they argue this approach has severely curtailed analysis of the complex interplay of human language, values, motives, cultures, reason, technologies and institutional contexts, which shape how construction processes unfold. For them, attending to these social interpretations necessarily involves suspending pre-given assumptions on behalf of the researcher about the ends of construction practitioners, such as 'quality, efficiency, productivity or profit' (Seymour and Rooke 1995, p. 522). Such a priori assumptions should, they argue, be suspended in order to register the full complexity of how social practices in construction are sustained. In this manoeuvre, Seymour and Rooke (1995) challenge the Platonic legacy of framing construction as a field to instrumentally apply technical and managerial knowledge to serve given technical and managerial ends. And yet, importantly, they do not challenge the performativity of knowledge per se: in a follow-up commentary, they explicitly state a requirement that CMR 'should enable practitioners to reflect upon their own practices in such a way as to facilitate their attempts to *improve* those practice' (Rooke et al. 1997, p. 494, emphasis added). They thus uphold the assumption that CMR knowledge is to be instrumental useful, although, importantly, they insert the caveat 'that usefulness [is] to be judged by the practitioners themselves' (Rooke et al. 1997, p. 494).

Seymour and Rooke's (1995) argument offers a valuable touchstone to understand how in recent decades some CMR has sought to rework construction as an object of knowledge. Since the publication of Seymour and Rooke's (1995) paper, there has been a notable growth in engagements with new areas of the social sciences and humanities in CMR—a 'social turn' perhaps. While much of this work does not follow the ethnomethodological approach of Seymour and Rooke (1995), it repeats the notion that the problems identified by the industry, now extended beyond a priori managerial, financial and technical concerns, can be better examined through the application of theories and methods from across the social sciences and humanities, rather than only with positivistic scientific approaches. Some of this work has been inspired by a single social thinker—for example, McCabe et al. (1998) employ Weber's theories of legitimization to diagnose the optimal qualities involved in instituting total quality management; Bresnen et al. (2005) draw on Gidden's structuration theory to examine conditions for the success of management change; and Ness (2010) engages Fairclough's Critical Discourse Analysis to evaluate the Respect for People HR agenda in UK construction. But, in the main, this body of work draws upon a specific theoretical approach, rather than single thinker, to address industry problems, such as Actor-Network Theory (ANT) to study design innovation (Harty 2008) and stakeholder complexity (Sage et al. 2011); practice theories to study inter-firm partnering (Bresnen 2009) and knowledge sharing (Styhre 2009); and ethnomethodology to study industry corruption (Rooke et al. 2004) and design coordination (Boudeau 2013).

This small sample of work cited here is merely indicative of the breadth of influence the social sciences and humanities—sociology, human geography, anthropology, philosophy and history—have had across dozens of studies published by scholars located in CMR groups and departments in CMR-orientated journals. Despite the evident breadth of CMR engagement with the social sciences and humanities, in most instances the 'social turn' in CMR demonstrates an asymmetry of engagement that mostly adheres to the theoretical instrumentalism found in Seymour and Rooke (1995). That is, while a growing number of CMR scholars make use of diverse theories and approaches from the social sciences and humanities to explore problems given in the industry, whether those of employers, employees, end-users, policymakers or trade unions, there remains a tendency throughout to apply social theories *for* construction rather than contribute to the development of such theories *with* construction. For example, McCabe et al. (1998) propose how they will argue with Weber 'that lack of senior management support deprives quality managers of the legal rational authority necessary to institute change … suggesting that a measure of charismatic authority constitutes a viable alternative' (p. 450). Despite developing such stimulating insights into the efficacy of charisma in post-bureaucratic work, McCabe et al. (1998) desist from discussing how they have developed/challenged/extended Weberian thinking. Similarly, Bresnen et al. (2005) apply the 'lens of structuration theory', in order to understand 'project management practices … as the outcome of a complex, and recursive, relationship between structural attributes and individual agency' (p. 551). While Bresnen et al. (2005) do consider contributions to theories of change in project-based organizations, they offer no contribution to structuration theory. Harty (2008) also evidences a similarly instrumental engagement with social theory in his much-cited analysis of construction innovation: 'An actor-network theory approach (ANT) is *used* to trace and unpack the interactions occurring around implementation of these artefacts' (p. 1029; emphasis added). Harty's (2008) notion of the relative boundedness of innovation, while clearly derived from ANT, is not engaged to inform wider ANT studies, but instead harnessed to challenge industry and scholarly assumptions about how and where innovation happens.

To summarize, within the 'social turn' in CMR, social theory appears largely engaged in decidedly instrumentalist terms. That this is the case is barely concealed: many authors explicitly declare how social theories are to be 'used' (Harty 2008, p. 1029; Walter and Styhre 2013, p. 1173), 'mobilized' (Ness 2010, p. 481), 'drawn upon' (Bresnen et al. 2005, p. 547) and thus can 'inform' (Styhre 2009, p. 997), be 'useful' (London and Pablo 2017, p. 554; Sage et al. 2011, p. 287; Tryggestad et al. 2010, p. 696), or form a 'lens' on the industry (Smiley et al. 2014, p. 805). Seldom are those same theories 'challenged', 'interrogated', 'critiqued' or 'problematized'. CMR scholars also often frame their engagement with theories from the social sciences and humanities in ways that lend the impression of scholars instrumentally selecting within an onto-epistemological toolkit to find the right means to view and explore previously neglected industry-given problem (see, e.g., Bresnen et al. 2005, pp. 550–551; McCabe et al. 1998, p. 448; Harty 2008, p. 1032), instead of a more sustained engagement with theories in terms of the empirical objects and problems they might form or not (for an exception, see Chan 2013, pp. 818–820).

To be clear, we do not wish to besmirch the value of applying social theory to explore problems facing the construction industry, or any other industrial sector, including those defined by managers, and indeed we have undertaken such work (e.g. Sage et al. 2010, 2011), but the tendency to apply rather than develop social theory does curtail the potential of CMR since the 'social turn' to reframe construction as a less instrumental object of knowledge. In developing this argument, we must also recognize that such CMR has (somewhat paradoxically) rehearsed the displacement of various Enlightenment binaries related to the instrumental framing of building construction: such as theory/practice with ethnomethodology (Seymour and Rooke 1995); designing/building (Tryggestad et al. 2010) and thought/matter (Sage et al. 2011) with ANT; signification/signified using Critical Discourse Analysis (Ness 2010); and mind/body using theories of aesthetic knowledge (Styhre 2009). But, crucially, these critical positions are mostly imported from other social science and humanities disciplines—CMR studies and scholars more seldom challenge and *develop* such disciplinary debates, whether within CMR-orientated journals or elsewhere.

This tendency seems underpinned by the preservation of the most deeply rooted binary framing building construction: the Platonic distinction between directive/calculative knowledge and the architectural depiction of building construction as the directive application of (technical, managerial, financial) knowledge. If theories from the social sciences and humanities are mostly used as tools to tackle problems manifest within the contemporary industry—even if those problems are not simply those of managers (although they mostly are for the reasons outlined by Lyotard 2002)—then that research will not only likely appeal less across the social sciences and humanities, but will also reinforce the separation of directive/calculative knowledge which underpins many of those other binaries and their social consequences.

We must recognize at this juncture that the boundaries of objects of knowledge are strongly institutionalized (Foucault 2004), serving specific interests and agendas. Scholars located in CMR-orientated research groups departments are certainly not as able to freely develop contributions to social theory as those in sociology, human geography or anthropology departments. Journal peer review processes, academic conferences, research associations (such as the Association of Research in Construction Management—ARCOM), editorial statements, published research, promotion processes and informal networks between peers, all serve the impression that:

> ... one of the crucial roles of social theory in construction research is to help break with taken-for-granted assumptions, thus creating the possibility for new policy and industry relevant insights into construction and contributing to the consolidation of construction research as a distinct field and to greater engagement with other social sciences .... (Schweber 2015, p. 840)

Through such discursive statements which specify and institutionalize objects of knowledge around social norms (Foucault 2004), the *application* of social theory to problems in the construction industry is valued as a, if not *the*, means around which the field of CMR is indeed to coalesce (see also Green and Schweber 2008). While some doubt that a field as theoretically diverse as CMR can be distinguished on the basis of the type of knowledge produced (e.g. Harty and Leiringer 2017), by briefly tracing the historical formation of construction as an object of knowledge across several centuries, CMR appears strongly shaped by the boundaries of 'directive', or performative (Lyotard 2002), knowledge which Plato demarcated for building over two thousand years ago (Plato 1961). Although some CMR scholars have called for wider contributions to 'discipline-rooted debates' (Leiringer and Dainty 2017, p. 1), the discursive formation of construction as a mostly instrumental object of knowledge appears widespread and deep-rooted and seems unlikely to change without also parallel shifts in other disciplines and fields about the social import of developing theories with construction. After all, the formation of construction in instrumental terms did not take place within CMR itself, or indeed within contemporary construction industries, but rather can be traced across the history of the humanities and social sciences—of Western thought. Consequently, if construction is going to be reworked as an object of knowledge more widely amenable to the social sciences and humanities, then the question must be asked how receptive are those disciplines to that shift in epistemic boundaries? It is this final question, and engagements with building construction outside of CMR, particularly within human geography, sociology and history, that we will now examine.

Construction Research Within the Social Sciences and Humanities

All social science and humanities disciplines, as perhaps all human thought, have a long-standing, though rather implicit, relationship with the people, places and politics of building construction, principally through their engagement with urbanizing societies. Urban life has long exerted an intellectual fascination for the disciplinary development of fields such as philosophy (Meagher 2008), literary studies (Fitzpatrick 2009; Jaye and Watts 1981), history (Ewen 2016), theology (Northcott 1998), sociology (Gottdiener and Hutchinson 2014; Lin and Mele 2013), human geography (Harvey 1973; Fyfe and Kenny 2005) and anthropology (Hannerz 1983; Jaffee and De Koning 2015). Of these disciplines, sociology, human geography and history are some of those most strongly concerned with urban studies, including through extensive studies of the built environment. It is for this reason that this edited collection engages explicitly with these disciplines as potentially those most receptive to stronger theorizing with building construction, while acknowledging significant cross-fertilizations with other social science and humanities disciplines, particularly anthropology, philosophy and politics, and also more interdisciplinary fields like management and organization studies.

Studies of urban life, and by extension the built environment, are closely linked to many of the most influential thinkers, and theories found across sociology, human geography and history. For instance, in sociology: Durkheim connected the shift towards organic solidarity, and an industrial division of labour to the increasing density of urban centres (Durkheim 1966); Weber (1978) discussed how urban growth in medieval Europe underpinned the urban emergence and diffusion of new forms of social domination such as rational bureaucracy; Tönnies (1963) famously associated urban growth with a shift from social relations orientated around instinctive will and communal solidarity (*Gemienschaft*) to those of capitalistic exchange, instrumental rationality and rational self-interest (*Gesselschaft*). Similarly, in human geography: Harvey (1973) predicated his call for a Marxist revolution in human geography, more attentive to the formation of socio-spatial inequalities, through the analysis of urbanism and its relationship with industrial capitalism; Smith (1986) examined urban renewal schemes to theorize how shifts and crises in the production of capital presage the arrival of new socio-spatial inequalities; and Massey (1994) developed her relational view of place with her own experiences of the global intersections of urban life in Kilburn, London. Equally, in history: Mumford's (1974) wide-ranging history documented how urbanization accompanied increases in social order, justice, moral stability, the accumulation of historical knowledge, as well as the proliferation of war, slavery, disease and labour overspecialization; Dyos (1973) explored the Victorian history of the London suburb of Camberwell to understand how suburbanization was enabled through a blend of population growth, technological developments in transport and expansions in housebuilding finance.

All of the above-cited thinkers have had a seminal influence over the development of the respective sub-disciplinary fields of urban sociology, urban geography and urban history. And yet, while the above scholars engage with changes in the built environment, it is notable that only one (Dyos 1973) addressed the lives of builders tasked with directly implementing those changes, although Dyos's (1973) accounts of the lives of the builders of Camberwell mostly relates to their activities as property developers.[4] Changes in the built environment—such as the expansion of suburban homes, highways, warehouse conversions, gated communities, walled cities (see Fyfe and Kenny 2005; Gottdiener and Hutchinson 2014) and now digital infrastructures (Riddlesden and Singleton 2014), eco-projects (Caprotti 2014) and micropubs (Hubbard 2017)—are undoubtedly a central analytical concern within (urban) history, sociology and geography. However, the experiences of the individuals responsible for building this urban architecture are seldom addressed (Clarke 1992). Or more precisely, only certain individuals—urban planners, developers and architects, along with the end-users of such spaces and places—feature prominently within urban sociologies, histories and geographies.

Further evidencing the absence of building construction, we can look to how three popular texts on urban life, spanning sociology, geography and history frame the boundaries of inquiry to the exclusion of the construction workforce. The first is taken from the introduction to *The City Cultures Reader* (Miles et al. 2000), where the editors (comprising a cultural theorist, a human geographer and an architectural historian) elaborate how the city can in part be understood in terms of the urban professionals that form it: 'Architects and urban designers ... [and] then there are the many other cultural designers and producers at work in the city: graphic designers, fashion designers, film makers, university lecturers, artists and musicians, writers and performers' (p. 2). Notably absent from this list, and from the readings with their text, are building

---

[4]Dyos (1973, p. 179) explains this decision as owing to the lack of cultural and social historical sources rather than as a reflection of his own lack of interest in the lives of builders. For histories of building, see, e.g., Clarke (1992), Jackson (1987), and Price (1980).

construction professionals: carpenters, bricklayers, steel-fitters, electricians and painters, quantity surveyors, planners, health and safety officers and site managers. The second similar exclusion appears in the popular urban sociology text—*The New Urban Sociology*. Summarizing their approach, Gottdiener and Hutchinson (2014) explain how 'settlement space is constructed and organized. It is built by people who have followed meaningful plan for the purpose of containing economic, political and cultural … metropolitan development is affected by government policy and by developers, financiers and other institutions in the real estate industry' (pp. 20–21). While emphasizing that urban life *is* built, they readily attribute building to the work of urban architects, planners and developers (see also Gottdiener and Hutchinson 2014, p. 19)—seemingly less concerned to recognize the experiences of those directly implementing such building. A third, similar, marginalization of builders can be found in Shane Ewen's (2016) *What Is Urban History?* The opening sentence of this text explicitly declares how urban history is not concerned with the 'city-building process' (Chapter 1, para. 1). Moreover, following the work of Dyos (1973), Ewen (2016) also frames engagement between urban historians and 'builders' solely in terms of their activities as financial property developers. Ewen (2016) also omits discussion of the two most sustained engagements with building labour and urban history (Clarke 1992; Price 1980).

Few scholars have attempted to explain this marginalization of building construction from urban studies. An exception here is Clarke (1992), who, introducing her building labour process analysis of the growth of London, argues that the physical obduracy of the built environment—'the structure imposed' (Clarke 1992, p. 18)—veils its social production by labour. Our difficulty with this 'building fetishism' critique is that it fails to address the specific agents of urban production that *are* addressed in urban studies (e.g. architects, urban planners, property developers) and also implies urban scholars can best (re)discover the experiences of builders with Marxist social theory, despite the long-standing influence of such approaches across the social sciences and humanities. Moreover, in many ways the building labour process is far more visible than many other forms of production due to its (usually) highly public setting. We contend that there is another, rather simpler, yet far older, reason for the tendency for urban geographers, historians and sociologists to sidestep the experience of builders. That is, if the work of builders, and by extension the building process, is merely understood as a physical, technical and financial 'instrument' for architects (and urban planners, policymakers and so on), as Alberti (1988, p. 3) put it over five centuries ago, then building construction inevitably appears largely passive to more the 'weighty' concerns of many urban sociologists, geographers and historians, such as suburbanization, gentrification and globalization. This would also help explain why builders only appear socially interesting when they play a financial or political role in urban development (as with Dyos 1973; Ewen 2016; Gottdiener and Hutchinson 2014). If this was all that could be said about geographical, sociological and historical interest in building construction, then our plea here for engagement with building construction would seem forlorn. Thankfully, a number of scholars, including many working from within urban geography, sociology and history, have, through a stream of more dedicated, and critical, engagements with the architecture and the built environment, started to develop a more sustained basis for engagement with construction by wrestling away the instrumentalist grip of Platonic and Albertian thinking.

Geographers have arguably been at the forefront of this attempt to displace the Albertian aggrandization of architecture and reciprocal debasements of building construction. Critical geographies of architecture (e.g. Lees 2001; Jacobs 2006; Rose et al. 2010; Lorne 2017) have increasingly challenged the notion that the architect is the epicentre of the meanings, uses and feelings of the built environment. Lees' (2001) ethnography of the design and use of Vancouver public library, for example, shows how embodied practices, such as homeless people washing in the toilets, enact architectural meanings of space in ways that are neither reducible to, or displacements of, the visions of architects (cf. Llewelyn 2003). Drawing on ANT and a discussion of the design and use of highrise apartments, Jacobs (2006) coins the term 'building event' to articulate the de-centring of architects, showing how human and non-human actors, from gas pipes to Le Corbusier, assemble the form and meaning of buildings (cf. Beauregard 2015; Kraftl 2010; Jacobs et al. 2007). Elaborating the affective register, Rose et al. (2010) study the use of an English shopping centre to explore how building events involve the assemblage of a milieu of human feelings as marble floors, spotlights, shopfronts, social interactions and past experiences interact, producing feelings of movement, belonging, safety, as well as frustration and sadness (cf. Adey and Kraftl 2008; Lees and Baxter 2011). Turning to politics, Lorne (2017) examines how the de-centring of architectural intent can be worked through new political imaginaries that allow 'design intelligence' to 'be deployed in open-ended, politically progressive ways without rejecting the skills of architects outright' (p. 276). Lorne (2017) offers a salient reminder that arguments around the de-centring of architectural intent are not purely theoretical concerns, rather such de-centring is also bound up with political struggles around the reduction of architecture to the task of realizing the most profitable building aesthetic (Gottschling 2017; Moran et al. 2016; Sage 2013). In challenging the aggrandization of architectural intent, and recognizing architecture as a relationally constituted process, this work opens up the potential to challenge the intellectual debasement of building construction as an instrument of architectural intent. Before discussing some recent geographical analyses of building construction responding to this potential, we will briefly consider the development of parallel arguments in sociology.

Critical sociological studies of architecture number far fewer than those in human geography, seemingly owing to the less prominent focus on the spatial environment within sociology. Nevertheless, some sociological studies (e.g. Gieryn 2002; Guggenheim 2009, 2013) offer a commensurate critique of neo-Albertian architectural thinking. One seminal contribution here is Gieryn's (2002) article *What Buildings Do*. Setting out the importance of buildings to 'stabilize social life', Gieryn (2002) laments—'It is surprising that they have been so rarely theorized by sociologists' (p. 35). Gieryn's (2002) thesis develops as a critique of how sociologists, specifically Giddens and Bourdieu, have missed the agency of the built environment—Giddens over-ascribing its production to human agency, Bourdieu to social structure. Influenced by ANT, and Science and Technology Studies more widely, Gieryn (2002), as Jacobs (2006), goes on to conceptualize 'building' as a heterogeneous, relational, achievement—a becoming through which new social worlds emerge (see also Müller and Reichmann 2015). Empirical analysis of Cornell University's Biotechnology Building grounds Gieryn's (2002) treatment of building, as he examines how design and use of the building shapes and is shaped by a particular way of doing science. Extending Gieryn's (2002) thesis further, Guggenheim (2009, 2013) argues how buildings are a particular type of socio-material assemblage: unlike other technologies (e.g. cars, guns, computers), their meanings and uses cannot be easily stabilized as they are strongly tied, over a long-time period, to a constantly changing environment; and their uses can be multiple *at the same time* (unlike, e.g., a computer, gun). However, they can, as Guggenheim (2013) explores via an analysis of changes in building use, coalesce, within certain thresholds, to assemble particular socio-material functions and meanings, such as banking and praying, as given in specific building types (e.g. banks, churches, libraries). Notwithstanding continued sociological arguments around the primacy of architectural intent, owing to their status in hegemonic socio-economic structures (e.g. Jones 2006), and the widespread focus in many sociological studies on building design and use over construction (as in Jones 2011; Müller and Reichmann 2015), sociologists, as geographers, appear increasingly receptive to the idea that the built environment is relationally constituted.

Taken together, these developments in sociology and human geography set the scene for a challenge to the long-standing idea, still commonplace with urban sociology, urban history and urban geography, that building construction is an instrument effecting the (largely technical, financial and physical) ends given by architects, developers, policymakers and end-users. Moreover, while we have focussed our review here on sociology and human geography, given the strength of arguments in those disciplines, the notion of the built environment as relationally constituted can also be found across the social sciences and humanities, spanning management and organization studies (e.g. Hirst and Humphreys 2013), anthropology (e.g. Allen 2014) and indeed history (e.g. Heynen 2013). It is thus perhaps unsurprising that across geography, human sociology and history, there has been an increasing interest in theorizing with building construction.

Before discussing recent work attending to building construction, we must first acknowledge that engagement within building construction within Anglophone social sciences and humanities is far from a recent invention. Certainly, interest in relational, materialist, ontologies, like ANT, is not a precondition for social studies of building. Five decades ago, Andrew Sykes (1969a, b) published two sociologies of social relations and work attitudes among iterant civil engineering workers in England. Six years later, Stewart Clegg's (1975) book-length study, *Power, Rule and Domination: A Critical and Empirical Understanding of Power in Sociological Theory and Organizational Life*, was published based on his ethnomethodology of organizational power in building sites in Northern England. The late 1970s and 1980s, also witnessed a series of industrial sociologies (e.g. Reimer 1979; Silver 1986) and anthropologies (e.g. Applebaum 1981, 1982) of construction in the USA, mostly based on ethnographic accounts. This research was accompanied by important histories of building work, such as Kenneth Jackson's (1987) *Crabgrass Frontier: The Suburbanization of the United States*, based partly on a history of suburbanization with American house builders, and two British histories of the emergence industrial capitalism with building labour from Richard Price (1980) in *Masters, Unions and Men: Work Control in Building and the Rise of Labour 1830–1914* and, more latterly, *Building Capitalism: Historical Change and the Labour Process in the Production of the Built Environment* written by Linda Clarke (1992).

Despite the breadth of these early contributions across the latter decades of the twentieth century, these engagements with building construction seemingly fell short of inspiring a sustained and coherent interest in building construction across the social sciences and humanities. This situation led authors, such as Clarke (1992) discussed above, to ponder at the lack of wider interest in building construction in (urban) sociology. There appear to be at least three explanations why early studies did not inspire a more coherent set of social studies of building compared to the early twenty first century. First, much of this early work took place under the banner of industrial sociology and anthropology—sub-disciplines that are now more or less subsumed under Business School fields such as human resource management and organization studies; this work thus became once step removed from sociology and anthropology departments (notably Clegg and Clarke have since spent most of their careers within business and management-based departments). Second, many of the above studies (e.g. Applebaum 1982; Reimer 1979; Silver 1986) focus markedly less on building theory *with* builders and rather more on mobilizing social theories to understand and address challenges within the sector, especially as related to poor industrial relations, the impact of management processes and building quality. In so doing, they effectively foreshadowed the instrumentalist orientation of the 'social turn' in CMR and are thus, we contend, hamstrung in engagement with the wider social sciences and humanities along similar lines. Third, these studies also remained largely disconnected within their disciplinary and national 'silos'—for example, Clarke (1992) only cites Price (1980), while Silver (1986) only cites the work of Applebaum (1982) and Reimer (1979). Thus, it is difficult to view these earlier mostly book-length studies as evidence of a coherent rising interest across the late twentieth-century social sciences and humanities in theorizing with building construction. To our mind, it is only as disciplines like sociology, human geography and history started to acknowledge and challenge the neo-Albertian instrumental framing of building construction, partly through the development of more relational approaches to architecture, that building theory *with* builders, and not purely *for*, builders could become a more coherent and sustained concern, including within a discipline lacking any early interest in building construction: human geography.

Within human geography, the rise of relational accounts of architecture from the turn of millennium (e.g. Lees 2001; Llewelyn 2003) has directly fed into steadily growing engagement with building construction as a place to develop geographical theories (Sage 2013). At the time of writing, this body of work encompasses studies of: landscaping with motorway construction (Merriman 2005); childhood as related to self-building school projects (Kraft 2006) and living on residential building sites (Kraft et al. 2013); national identities with polish builders (Datta 2008); urban maintenance with concierge workers in a high-rise housing blocks (Jacobs et al. 2012; Strebel 2011); urban place-making with the repair of a Manchester church (Edensor 2011); professional identities with English commercial construction (Sage 2013); geographies of incarceration and rehabilitation with Building Information Modelling (BIM) and prison construction (Moran et al. 2016); architectural politics with construction procurement routes (Gottschling 2017); and home-making with new residents living on Australian housebuilding sites (Gillon 2017). These diverse geographical engagements with building construction have all been influenced by relational approaches to architecture, and in particular ANT. The convergence of ANT and more sustained theorizing of building construction is also vividly apparent in other fields, such as management and organizational studies where early proponents of ANT developed their thinking with building construction (e.g. Gherardi and Nicolini 2000; Suchman 2000). In order to understand the confluence of building construction and ANT, which has been noted here repeatedly and is evident in three chapters within this volume, we can turn to Latour:

> Usually the great advantage of visiting construction sites is that they offer an ideal vantage point to witness the connections between humans and non-humans. Once visitors have their feet deep in the mud, they are easily struck by the spectacle of all the participants working hard at the time of their most radical metamorphosis … Even more important, when you are guided to any construction site you are experiencing the troubling and exhilarating feeling that things could be different, or at least that they could still fail – a feeling never so deep when faced with final product, no matter how beautiful or impressive it may be. (pp. 88–89)

This passage from Latour (2005) demonstrates how the combination of ANT and building construction offers a potent mix to displace long-standing binaries, such as mind/body, human/non-human, design/execution, directive/calculative knowledge, substantive/instrumental reason, theory/practice and subject/object—binaries which have, at least since Plato, been powerfully reproduced through the discursive formation of building construction. However, while the influence of ANT, and indeed Latour, within human geography, and some management and organization studies of building construction remains strong (see, e.g., Tryggestad and Georg 2011; Sage et al. 2013), recent sociological studies of building construction are distinctly less Latourian—despite calls from some architectural sociologists (as in Gieryn 2002; Guggenheim 2009, 2013). Instead, construction studies published by scholars based in sociology departments and research groups (e.g. Ajslev et al. 2016; Powell and Sang 2015; Thiel 2007, 2010, 2013; Watts 2007) more closely resemble older industrial and urban sociologies of building (e.g. Clegg 1975; Reimer 1979; Silver 1986). In developing his ethnographic sociologies of builders, Thiel (2007, 2010, 2013), for example, enters into dialogue with Marx and Bourdieu, not Latour, Callon or Deleuze, as he discusses the reproduction of class- and gender-bounded identities in English building sites. As such, Thiel's (2007, 2010, 2013) work offers somewhat of a rejoinder to the emphasis on the dissolution of Enlightenment binary thinking around construction as encouraged by ANT. As Thiel (2007) explains:

> In the organization of building work, abstract knowledge became conceptually separate from execution: architects, surveyors and building site managers planned works, and labourers and tradesmen physically constructed them. As a result, the builders drew a distinction between themselves and the management, and this divide was expressed as 'being in the office' or 'being on the tools' … (p. 237) Rhetorics about working hard, drinking hard, fighting hard, and fucking hard were a dominant scaffold that underpinned and infused the builders' *public* culture. (p. 241)

Thiel's (2007, 2010, 2013) studies of construction document how class- and gender-bounded identities effect boundaries around the career trajectories of builders. Here, Platonic–Albertian separations between architecture/construction, clean/dirty work (and theory/practice, mind/ body, calculative/directive knowledge and so on) are figured as more than intellectual positions (as with Latourian studies)—they uphold a division of labour and accompanying inequalities. Emphasizing the historicality of such processes, Thiel (2007) further notes how working-class cultures in construction appear articulated around 'an almost feudal valorisation of strength and protection' (p. 246), which when viewed alongside the relatively autonomy of building work from industrial management (Thiel here cites the work of Applebaum 1981, 1982), indicates pre-industrial influences on contemporary working-class cultures.

Other sociological studies of building construction (e.g. Powell and Sang 2015; Ajslev et al. 2016; Powell and Sang 2015; Watts 2007) have offered similar expositions of class and gender with building construction. Watts (2007: 309), studying the experiences of women working as civil engineers, discusses how the public denial of the social significance of construction (as ushered by Platonic and Albertian thinking) delimits their professional pride in construction is coded as male/masculine 'dour' technical work, serving to inhibit gender diversity and inclusivity in the sector. Elaborating further on the persistent reproduction of construction as masculine, male-dominated, work, Powell and Sang (2015) explore, via Bourdieu, how women construction engineering students (and architects) misrecognize discrimination as part of the natural order of things—whereby, for example, men are said to be naturally better at physical, technical work—rather than as symbolic violence often knowingly undertaken against them by men. Similarly, in a study of Scandinavian site-based construction workers, Ajslev et al. (2016) examine how work-related physical pain and stress are rationalized as a product to be exchanged for a wage within a working-class, masculine work environment. Further critical treatments of how binaries such as mind/body, design/execution and calculative/directive knowledge remain bound up in gender- and class-based social identities can be found within other sociologies of construction (e.g. Eisenberg 1998b; Paap 2006; Wright 2016), construction studies published within sociologically inclined management and organizational studies journal (see, e.g., Fletcher and Watson 2007; Ness 2012; Parker 2016), as well as within construction anthropologies (Yarrow and Jones 2014) and histories (Heynen 2013).

Having surveyed some of the key disciplinary gathering points, bifurcations, boundaries and directions of travel, in existing engagements with building construction across sociology, human geography and history, we will consider the unrealized potential in this body of work. These shortcomings relate not so much to the content of individual studies, as to the lack of reflection on this body of work as a whole. And, in particular, the possibility that this work can become more than simply the sum of its rather disconnected parts; it is this possibility that underpins the purpose of this book.

Summary Discussion

Before introducing the following chapters, we will briefly reflect again upon the purpose of the present volume: to offer a sustained acknowledgement, understanding and challenge to the marginalization of construction as a place to develop novel understandings of society. By elaborating upon the discursive formation of construction as an object of knowledge and then tracing the effects of this discursive formation across the 'social turn' in CMR and its reworking across the social sciences and humanities, we have endeavoured across this introduction to go some way in addressing this aim. Specifically, we have sought to show that the marginalization of building construction as a place to understand society is neither a recent invention and nor is it absolute; yet it is only relatively recently that the origins and consequences of this intellectual debasement have started to be more fully understood and challenged. Our review of recent engagements with building construction throws into relief that the strength of this challenge remains somewhat hamstrung by the relative detachment of recent debates, suggestive of the unrealized potential to develop a more consolidated set of interdisciplinary enquiries, as this volume seeks to offer, around the social studies of building work.

We can start to better appreciate this potential by reflecting upon work in human geography and sociology. Recent sociological research on construction can be distinguished from geographical work on the basis that the former is notably less concerned with displacing the use of the construction to effect long-standing binaries (around designing/making, mind/body, theory/practice, thought/matter, object/subject, calculative/directive knowledge, etc.) and is rather more concerned with tracing how the cultural and social framing of construction, around those binaries, reproduces lived social inequalities. Consequently, it would seem to us that these two bodies of work are entirely complementary. This is because while it is clearly important to recognize the social significance of building construction, as geographers have, by challenging its intellectual circumscription by the binary thinking exemplified by Plato, Vitruvius, Alberti and Laugier, it is equally important, as with sociological studies, to recognize the prevailing consequences of those social and cultural framings as integral to that overlooked social significance. The social significance of building construction cannot simply be rediscovered through Latourian analyses of the 'building event' (e.g. Jacobs 2006; Gottschling 2017), as this would deny that attempts to detach building construction from society, as through myths of architectural intent that render building work an unthinking, instrumental, activity were not in themselves socially consequential and thus form part of that (overlooked) social significance. But, by the same token, neither can the social significance of building work solely correspond to sociological and historical studies of the consequences of the social and cultural framings of construction (e.g. Ajslev et al. 2016; Powell and Sang 2015; Thiel 2007, 2010, 2013; Watts 2007), while the ontologies and epistemologies that underpin those framings (e.g. Edensor 2011; Merriman 2005; Kraft 2006; Jacobs et al. 2012; Sage 2013; Strebel 2011) are left unquestioned, and largely intact. What is more, it is notable that both of these recent engagements with building construction remain mostly disconnected from wider work within those respective disciplines that concern the intersection of society and the built environment, especially within urban history, urban sociology and urban geography. And many of these debates have also developed entirely separately from the 'social turn' in CMR. As a result, some of the most important intellectual conduits that uphold the instrumentalization of building construction—construction management and engineering and urban studies—remain more or less untouched by recent reworkings of construction as an object of knowledge in the social sciences and humanities. This state of affairs will likely both delimit the force of those reworkings themselves and also circumscribe the potential for scholars working within construction management and urban studies to recognize the potency of building construction to develop new theorizations of societies, politics, geographies, ethics, identities, histories and futures. This is not merely a scholarly concern either. Many of the most pressing challenges facing human societies in the early twenty-first century, such as climate change, ageing societies, urban growth, the proliferation of weapons of mass destruction and social inequality, have advanced through a framing of building construction as a means serving narrowly managerial, technical and financial ends. Changing this framing of construction—reforming construction as an object of knowing—would thus seem essential to tackle those challenges.

By bringing together scholars working from across sociology, human geography and history departments, as well as those working in the 'social turn' in CMR and interdisciplinary fields such as management and organization studies, the purpose of this volume is to encourage a more sustained reflection on the potential to develop such novel theorizations engagements to challenge the social marginalization of building construction. This challenge will not be quickly won. After all, the formation and institutionalization of construction as a debased object of knowledge for social science and humanities inquiry has taken centuries, and is, as our review here attests, evidently still at work. Our hope is that this interdisciplinary volume offers readers from across the social sciences and humanities, and within construction management and engineering, new concepts, methodologies and empirical directions to understand and challenge this situation and its consequences.

Summary of Chapters

We have organized the chapters within this volume around a transition between different disciplinary *foci*, starting with history, through sociology and onto human geography. In making these disciplinary cuts, we were not so much guided by the institutional affiliation of the contributing authors, indeed many of the authors below have moved across multiple disciplinary 'homes', but rather the intellectual focus of the contribution within each chapter.

One of the most problematic legacies of Albertian binaries between design/execution, architecture/construction, theory/practice and mind/body is the idea that buildings possess a static and singular form and function. Through a historical study of adaptations to a university library, Patel and Tutt directly challenge such notions. But in so doing, they also argue against the commonplace alternative idea that buildings are relationally constituted black boxes (Jacobs 2006; Gieryn 2002), or they congeal, however temporary, into a single building type (Guggenheim 2013). Instead through the development of an innovative methodology—combining archival analysis, ethnographic observations, studies of the buildings itself and interviews—the authors reflect on the lived multiplicity of building as they interact with other bodies, including their own. As such, they also respond to Heynen's (2013) call for historical architectural studies that go beyond the symbolic to consider bodily and material registers (cf. Lees 2001). Their account of the University of Reading Library, from its origin through refurbishments to recent events, including an exhibition of its past, considers how discussions of multiplicity, performativity and fluidity with ANT thinking can contribute to social studies of building work, while also discussing some of the challenges that come with understanding building construction as always multiple, always incomplete.

Inspired by earlier construction histories, such as Clarke (1992), which challenged architectural and urban historians to address the production of cities by building labour, Paskins's chapter develops a labour history of building sites in the Paris region during the post-war construction 'boom' of the 1960s. He explains how the current inattention to building construction within contemporary urban (history) studies echoes long-standing marginalizations of the 'building site' by architects, policymakers and the media, not least those in Paris in the 1960s. As he argues, these enduring intellectual, political and cultural erasures of building construction inevitably serve political ideologies that profit, politically, economically and socially, from subjugating the lives of construction workers. Developing a counter to these dominant framings, Paskins shows, with reference to a wide range of historical sources, how changes in building practices were bound up in national debates around immigration, identity and workers' rights. He thus opens an invitation for urban studies scholars to consider how histories of building construction can develop new understandings of the politics of cities and nations.

As a male-dominated work environment, social studies of building have frequently contributed to the theorization of gender norms in male-dominated workplaces (e.g. Chan 2013; Ness 2012; Powell and Sang 2015; Smith 2013; Thiel 2007, 2013; Watts 2007, 2009; Wright 2016). Drawing upon an ethnography of the embodied experiences of a single female site manager working within the Scandinavian construction industry Sandberg, Räisänen, Löwstedt and Raiden challenge the emphasis on gender norms, and their disciplining effects, in earlier studies of construction (e.g. Powell et al. 2009; Smith 2013), by considering how social studies of building can also contribute to debates in gender studies on resistance and intersectionality. Regarding the former, the experiences of their female participant are mobilized to show that women in male-dominated workplaces can actively co-opt and rework gender norms to their benefit to a far greater extent than extant work on construction, and other industries, usually acknowledges. In a further challenge to previous social studies of building construction and gender, which tend to focus exclusively on gender, they also contribute to debates around intersectionality. Specifically, they argue that female (and male) bodies are not simply, or indeed predominately, subjugated by gender norms, indeed these are seldom acknowledged by their female participant, but rather the scars—mental and physical—construction workers experience, mostly correspond to the intensification of work by fast-moving speculative capital. And moreover, they propose it is class-bound identities, not simply those gender, that may offer the most useful discursive resources in navigating, accommodating and resisting these challenges.

The next contribution to theory building with building concerns institutional theory. Tracing its origins at least as far as Weberian sociology (Weber 1978), recent sociological theories of institutions (often referred to as 'neo-institutionalism') are now part of the intellectual furniture of sociological studies, especially those concerning organizational life. Explicating how social rules, logics and processes shape human behaviour, sociological institutional theories present a powerful alternative to rational economic models and cognitive approaches in understanding human behaviour. However, while many institutional theorists have increasingly turned away from understandings that privilege stability and continuity towards considering how flexibility and change happen in processes of institutionalization, CMR scholars working with institutional theories tend to remain strongly gripped by processes of stabilization such that processes of change are left more or less unexamined, or appear as a punctuated series of episodes of institutionalization and de-institutionalization. In his review-based contribution, Chan registers the established tendency for construction scholars to import and apply rather than engage and contribute to sociological debates. Challenging this situation in CMR scholarship, Chan considers how the fast-paced context and the specific materialities of construction offers a fertile ground to contribute new ideas around change and change management in institutional theories. Developing this line of thinking, he identifies three avenues for inquiry along which social studies of building might contribute new understandings of institutional change. First, the construction industry, because of its project-based nature, relies upon routines that encompass both shorter-term exploration and longer-term survival and is thus a rich setting to study how every institutionalized routine involves a degree of de-institutionalization, revealing how change is interdependent with stability. Second, the construction industry, where the history of pre-industrial medieval craftwork presages a certain degree of work autonomy (Ingold 2013; Sennett 2008; Thiel 2013), provides a wealth of empirical opportunities to challenge the emphasis on elite actors in recent studies of institutional change. Finally, the construction industry, which in most industrial definitions also encompasses process of demolition (e.g. ISIC 2017), is rich with stories concerning what can be gained from processes of demise and ruin and can thus contribute to recent debates in institutional theory on the productivity of de-institutionalization.

Contributing to geographical scholarship, Grubbauer proposes that social studies of building look beyond the Global North—where building may at least appear more sporadic, more containable, less noticeable perhaps—to also consider places where building is more continuous, such as the informal, self-built, settlements of the *colonias populares* (people's neighbourhood) of Mexico City. Grubbauer is specifically concerned with understanding the increasingly formal marketization and internationalization of informal urban growth in the Global South. To explore these changes in self-building, she examines a set of policies by the Mexican government in recent decades to improve the housing of the *colonias populares* by providing self-build homes with legal recognition and offering microfinance, subsidies and technical assistance to self-builders. As she explains, many of these changes have accompanied and indeed brought about an increasing reliance upon national and global construction suppliers—resulting in a shift away from local wood and brick construction towards concrete blocks. Through her study of self-help building, Grubbauer explores the social and political geographies of these changes, acknowledging that while the self-building of *colonias populares*, and similar places elsewhere in Latin America and the Global

South, is increasingly bound up with imperatives for profit-maximization in banking and the global construction industry, it is not, as might be expected, simply another form of (neoliberal) capitalist exploitation and cultural imperialism. This is because, as Grubbauer argues, these changes also foster increases in the self-worth of individuals which are necessary for more collective political agency to enable lower-income communities to advance their interests. Grubbauer's study demonstrates the potential for social studies of building to generate new contributions to important debates within urban studies concerning the shifting material and symbolic politics of urban development in the Global South.

Continuing to elaborate on the theme of the geographies of building, as related to place-making and housebuilding, Sage and Vitry develop a case study of the pre-construction development and planning phase of a large housebuilding project in the English midlands to expound an ANT theory of place. As they note, although ANT has often been engaged to understand space and scale, especially within human geography, it has seldom informed theories of place. Though an analysis examining how a consortium of housebuilding firms and a protest group worked with five different simplifications of a single place in order to negotiate their different political interests and agendas with the local planning authority, they develop a more place-sensitive version of ANT. As with Patel and Tutt, their study focusses less on the dynamics of construction sites and more on the wider construction industry. Empirically, their chapter demonstrates the wider involvement that construction firms, as opposed to architects and urban planners, have within processes of urban development and place-making. More theoretically, the ANT theory of place they propose challenges recent relational theories of place in geography to consider place-making as regressive, future orientated and affective.

In the final contribution to the collection, Sage, Justesen, Dainty, Tryggestad and Mouritsen challenge the prevailing emphasis on people and technologies in social studies of building by analysing relationships between humans and animals, and their respective building processes and habitats. Their study engages with construction sites to theorize the role of heterogeneous agencies within processes of human and animal organizing, with a particular attention to spacings and timings of organization. Using two case studies of construction projects in the UK and Scandinavia, they argue that non-human animals and their habitats are, when legally protected, far from passively dominated by human processes of organizing, such as those of construction management, as critical animal studies sometimes suggest, but can also be rather more actively implicated in human organizing. As a result, social studies of building are shown to offer a rich potential to inform debates across the social sciences and humanities regarding the entanglement of humans with other forms of life.

References

Adey, P., & Kraftl, P. (2008). Architecture/affect/inhabitation: Geographies of being-in buildings. *Annals of the Association of American Geographers, 98*(1), 213–231.

Ajslev, J., Møller, J., Persson, R., & Andersen, L. (2016). Trading health for money: Agential struggles in the (re)configuration of subjectivity, the body and pain among construction workers. *Work, Employment & Society*, 1–17. https://doi.org/10.1177/0950017016668141.

Alberti, L. (1450/1988). On the art of building (J. Rykwert, N. Leacy, & R. Tavernor, Trans.). Cambridge, MA: The MIT Press.

Allen, S. (2014). An award controversy: Anthropology, architecture, and the robustness of knowledge. *Journal of Material Culture, 19*(2), 169–184.

Alvesson, M., & Wilmott, H. (Eds.). (1993). *Critical management studies.* London: Sage.

Alvesson, M., Bridgman, T., & Wilmott, H. (Eds.). (2009). *The Oxford handbook of critical management studies.* Oxford: Oxford University Press.

Applebaum, H. (1981). *Royal blue: The culture of construction workers.* London: Holt, Rinehart and Winston.

Applebaum, H. (1982). Construction management: Traditional versus bureaucratic methods. *Anthropological Quarterly, 55*(4), 224–234.

Baddo, N. (2008). Stainless steel in construction: A review of research, applications, challenges and opportunities. *Journal of Constructional Steel Research, 64,* 1199–1206.

Battisti, E. (1981). *Brunelleschi: The complete work.* London: Thames and Hudson.

Beauregard, R. (2015). We blame the building! The architecture of distributed responsibility. *International Journal of Urban and Regional Research, 39*(3), 533–549.

Boudeau, C. (2013). Design team meetings and the coordination of expertise: The roof garden of a hospital. *Construction Management and Economics, 31*(1), 78–89.

Bourdieu, P. (1984). *Distinction.* London: Routledge.

Bresnen, M. (2009). Living the dream? Understanding partnering as emergent practice. *Construction Management and Economics, 27*(10), 923–933.

Bresnen, M. (2017). Being careful what we wish for? Challenges and opportunities afforded through engagement with business and management research. *Construction Management and Economics, 35*(1–2), 24–34.

Bresnen, M., Goussevskaia, A., & Swan, J. (2005). Implementing change in construction project organizations: Exploring the interplay between structure and agency. *Building Research & Information, 33*(6), 547–560.

Caprotti, F. (2014). Eco-urbanism and the eco-city, or, denying the right to the city? *Antipode, 46*(5), 1285–1303.

Chan, P. (2013). Queer eye on a 'straight' life: Deconstructing masculinities in construction. *Construction Management and Economics, 31*(8), 816–831.

Cheah, C., Kang, J., & Chew, D. (2007). Strategic analysis of large local construction firms in China. *Construction Management and Economics, 25*(1), 25–38.

Clarke, L. (1992). *Building capitalism: Historical change and the labour process in the production of the built environment.* London: Routledge.

Clarke, L., Winch, C., & Brockman, M. (2013). Trade-based skills versus occupational capacity: The example of bricklaying in Europe. *Work, Employment & Society, 27*(6), 932–951.

Clegg, S. (1975). *Power, rule and domination: A critical and empirical understanding of power in sociological theory and organization life.* London: Routledge and Kegan Paul.

Dainty, A. (2008). Methodological pluralism in construction management research. In A. Knight & L. Ruddock (Eds.), *Advanced research methods in the built environment* (pp. 1–13). Oxford: Wiley.

Datta, A. (2008). Building differences: Material geographies of home(s) among Polish builders in London. *Transactions of the Institute of British Geographers, 33,* 518–531.

Durkheim. (1966). *The division of labour in society.* New York: Free Press.

Dyos, H. G. (1973). *Victorian suburb: A study of the growth of Camberwell.* Leicester: Leicester University Press.

Edensor, T. (2011). Entangled agencies, material networks and repair in a building assemblage: The mutable stone of St Ann's Church, Manchester. *Transactions of the Institute of British Geographers, 36,* 238–252.

Eisenberg, S. (1998a). *Pioneering: Poems from the construction site.* Ithaca, NY: Cornell University Press.

Eisenberg, S. (1998b). *We'll call you if we need you: Experiences of women working in construction.* Ithaca, NY: Cornell University Press.

Ewen, S. (2016). *What is urban history?* Cambridge: Polity Press.

Feenberg, A. (1999). *Questioning technology.* London: Routledge.

Fitzpatrick, J. (2009). *The idea of the city.* Newcastle: Cambridge Scholars Publishing.

Fletcher, D., & Watson, T. (2007). Voice, silence and the business of construction: Loud and quiet voices in the construction of personal, organizational and social realities. *Organization, 14*(2), 155–174.

Foucault, M. (2004). *The archaeology of knowledge.* London: Routledge.

Fournier, V., & Grey, C. (2000). At the critical moment: Conditions and prospects for critical management studies. *Human Relations, 53*(1), 7–32.

Fyfe, N., & Kenny, J. (Eds.). (2005). *The urban geography reader.* London: Routledge.

Gherardi, S., & Nicolini, D. (2000). To transfer is to transform: The circulation of safety knowledge. *Organization, 7*(2), 329–348.

Gieryn, T. (2002). What do buildings do? *Theory and Society, 31,* 35–74.

Gillon, C. (2017). Under construction: How home-making and underlying purchase motivations surface in a housing building site. *Housing, Theory and Society.* https://doi.org/10.1080/14036096.2017.1337650.

Glasser, M. (2008). *Construction site: Metamorphosis in the city.* Zurich: Lars Müller Publishers.

Gottdiener, M., Huthchinson, R., & Ryan, M. (2014). *The new urban sociology* (5th ed.). Boulder, CO: Westview Press.

Gottschling, P. (2017). Architectural scripts: Construction procurement within the user worlds of building events. *Social and Cultural Geography.* https://doi.org/10.1080/14649365.2017.1296178.

Green, S. D., & Schweber, L. (2008). Forum theorizing in the context of professional practice: The case for middle-range theories. *Building Research & Information, 36*(6), 649–654.

Gregory, D. (1994). *Geographical imaginations.* Oxford: Blackwell.

Guggenheim, M. (2009). Building memory: Architecture, networks and users. *Memory Studies, 2*(1), 39–53.

Guggenheim, M. (2013). Unifying and decomposing building types: How to analyse the change of use of sacred buildings. *Qualitative Sociology, 36,* 445–464.

Hannerz, U. (1983). *Exploring the city: Towards an urban anthropology.* New York City, NY: Columbia University Press.

Harriss, C. (1998). Why research without theory is not research. A reply to Seymour, Crook and Rooke. *Construction Management and Economics, 16*(1), 113–116.

Harty, C. (2008). Implementing innovation in construction: Contexts, relative boundedness and actor-network theory. *Construction Management and Economics, 26*(10), 1029–1041.

Harty, C., & Leiringer, R. (2017). The futures of construction management research. *Construction Management and Economics, 35*(7), 392–403.

Harvey, D. (1973). *Social justice and the city.* London: Edward Arnold.

Harvey, D. (1989). *The condition of postmodernity.* Oxford: Blackwell.

Hayes, N. (2002). Did manual workers want industrial welfare? Canteens, latrines and masculinity on British building sites 1918–1970. *Journal of Social History, 35*(3), 637–658.

Heynen, H. (2013). Space as receptor, instrument or stage: Notes on the interaction between spatial and social constellations. *International Planning Studies, 18*(3–4), 342–357.

Hirst, A., & Humphreys, M. (2013). Putting power in its place: The centrality of edgelands. *Organization Studies, 34*(10), 1505–1527.

Hubbard, P. (2017). Enthusiasm, craft and authenticity on the High Street: Micropubs as 'community fixers'. *Social and Cultural Geography.* https://doi.org/10.1080/14649365.2017.1380221.

Ingold, T. (2013). *Making: Anthropology, archaeology, art and architecture.* London: Routledge.

ISIC. (2017a). International Standard Industrial Classifications Revision 4, United Nations Statistics Division. Retrieved from https://unstats.un.org/unsd/cr/registry/regcst.asp?Cl=27.

ISIC. (2017b). International Standard Industrial Classifications Revision 4—Section F Construction, United Nations Statistics Division. Retrieved from https://unstats.un.org/unsd/cr/registry/regcs.asp?Cl=27&Lg=1&Co=F.

Jackson, K. T. (1987). *Crabgrass frontier: The suburbanization of the United States.* Oxford: Oxford University Press.

Jacobs, J. (2006). A geography of big things. *Cultural Geography, 13,* 1–27.

Jacobs, J., Cairns, S., & Strebel, I. (2007). 'A tall storey … but, a fact just the same': The red road high-rise as a black box. *Urban Studies, 44*(3), 609–629.

Jacobs, J., Cairns, S., & Strebel, I. (2012). Doing building work: Methods at the interface of geography and architecture. *Geographical Research, 50*(2), 126–140.

Jaffee, R., & De Koning, A. (2015). *Introducing urban anthropology*. London: Routledge.

Jaye, M., & Watts, A. (Eds.). (1981). *Literature and the American urban experience*. Manchester: Manchester University Press.

Jones, P. (2006). The sociology of architecture and the politics of building: The discursive construction of Ground Zero. *Sociology, 40*(3), 549–565.

Jones, P. (2011). *The sociology of architecture*. Liverpool: Liverpool University Press.

King, R. (2000). *How a renaissance genius reinvented architecture*. London: Penguin.

Klikauer, T. (2013). *Managerialism*. Basingstoke: Palgrave Macmillan.

Kraftl, P. (2006). Building an idea: The material construction of an ideal childhood. *Transactions of the Institute of British Geographers, 31*, 488–504.

Kraftl, P. (2010). Architectural movements, utopian moments: (In)coherent renderings of the Hundertwasser-Haus, Vienna. *Geografiska Annaler: Series B, 92*, 327–345.

Kraftl, P., Christensen, P., Horton, J., & Hadfield-Hill, S. (2013). Living on a building site: Young people's experiences of emerging 'sustainable communities' in England. *Geoforum, 50*, 191–199.

Lahdenpera, P. (2012). Making sense of the multi-party contractual arrangements of project partnering, project alliancing and integrated project delivery. *Construction Management and Economics, 30*(1), 57–79.

Latour, B. (2005). *Reassembling the social: An introduction to actor-network theory*. Oxford: Oxford University Press.

Laugier, M. (1753/1977). *An essay on architecture*. Los Angeles: Hennessey Ingallis.

Lees, L. (2001). Towards a critical geography of architecture: The case of an ersatz colosseum. *Cultural Geographies, 8*(1), 51–86.

Lees, L., & Baxter, R. (2011). A 'building event' of fear: Thinking through the geography of architecture. *Social and Cultural Geography, 12*(2), 107–122.

Leiringer, R., & Dainty, A. (2017). Construction management and economics: New directions. *Construction Management and Economics, 35*(1–2), 1–3.

Lin, J., & Mele, C. (Eds.). (2013). *The urban sociology reader* (2nd ed.). London: Routledge.

Llewelyn, M. (2003). Polyvocalism and the public: 'Doing' a critical historical geography of architecture. *Area, 35*(3), 264–270.

London, K., & Pablo, Z. (2017). An actor–network theory approach to developing an expanded conceptualization of collaboration in industrialized building housing construction. *Construction Management and Economics, 35*(8–9), 553–577.

Lorne, C. (2017). Spatial agency and practising architecture beyond buildings. *Social and Cultural Geography, 18*(2), 268–287.

Lyotard, J.-F. (2002). *The postmodern condition: A report on knowledge* (G. Bennington & B. Massumi, Trans.). Minneapolis: University of Minnesota Press.

Massey, D. (1994). *Space, place and gender.* Minneapolis: University of Minnesota Press.

Meagher, S. (2008). *Philosophy and the city.* Albany, NY: Suny Press.

McCabe, S., Rooke, R., Seymour, D., & Brown, P. (1998). Quality managers, authority and leadership. *Construction Management and Economics, 16*(4), 447–457.

McCann, D., & Forde, M. (2001). Review of NDT methods in the assessment of concrete and masonry structures. *NDT & E International, 34,* 71–84.

Merriman, P. (2005). 'Operation motorway': Landscapes of construction on England's M1 motorway. *Journal of Historical Geography, 31,* 113–133.

Moran, D., Turner, J., & Jewkes, Y. (2016). Becoming big things: Building events and the architectural geographies of incarceration in England and Wales. *Transactions of the Institute of British Geographers, 41,* 416–428.

Miles, M., Hall, T., & Borden, I. (Eds.). (2000). *The city cultures reader.* London: Routledge.

Müller, A., & Reichmann, M. (2015). *Architecture, materiality and society: Connecting sociology of architecture with science and technology studies.* Basingstoke: Palgrave Macmillan.

Mumford, L. (1974). *The city in history.* Harmondsworth: Pelican Books.

Ness, K. (2010). The discourse of 'respect for people' in UK construction. *Construction Management and Economics, 28*(5), 481–493.

Ness, K. (2012). Constructing masculinity in the building trades: 'Most jobs in the construction industry can be done by women'. *Gender, Work and Organization, 19*(6), 654–676.

Northcott, M. (Ed.). (1998). *Urban theology: A reader.* New York, NY: Bloomsbury.

ONS. (2017). SIC Definition of the Construction Industry, Office of National Statistics. Retrieved from https://www.ons.gov.uk/ons/rel/construction/construction-statistics/no--16--2015-edition/pdf-construction-statistics-appendix-2.pdf.

Paap, K. (2006). *Working construction—Why white working-class men put themselves—And the labour movement—In harm's way*. Ithaca, NY: Cornell University Press.

Parker, M. (2002). *Against management*. Cambridge: Polity Press.

Parker, M. (2016). Tower cranes and organization studies. *Organization Studies*. https://doi.org/10.1177/0170840616663246.

Plato. (360 BCE/1961). *The Statesman* (J. B. Skemp, Trans.). London: Routledge.

Pont, G. (2005). The education of the classical architect from Plato to Vitruvius. *Nexus Network Journal, 7*(1), 75–85.

Powell, A., & Sang, K. (2015). Everyday experiences of sexism in male-dominated professions: A Bourdieusian perspective. *Sociology, 49*(5), 919–936.

Powell, A., Bagilhole, B., & Dainty, A. (2009). How women engineers do and undo gender: Consequences for gender equality. *Gender, Work and Organization, 16*(4), 411–428.

Price, R. (1980). *Masters unions and men: Work control in building and the rise of labour 1830–1914*. Cambridge: Cambridge University Press.

Purdy, D. (2011). *On the ruins of Babel*. Ithaca, NY: Cornell University Press.

Raftery, J., McGeorge, D., & Walters, M. (1997). Breaking up methodological monopolies: A multi-paradigm approach to construction management research. *Construction Management and Economics, 15*(3), 291–297.

Riddlesden, D., & Singleton, A. (2014). Broadband speed equity: A new digital divide? *Applied Geography, 52,* 25–33.

Riemer, J. (1979). *Hard hats: The work world of construction workers*. Beverly Hills, CA: Sage.

Roberts, T. (2012). From 'new materialism' to 'machinic assemblage': Agency and affect in IKEA. *Environment and Planning D: Society and Space, 44,* 2512–2529.

Rooke, J., Seymour, D., & Crook, D. (1997). Preserving methodological consistency: A reply to Raftery, McGeorge and Walters. *Construction Management and Economics, 15*(5), 491–494.

Rooke, J., Seymour, D., & Fellows, R. (2004). Planning for claims: An ethnography of industry culture. *Construction Management and Economics, 22*(6), 655–662.

Rose, G., Degen, M., & Basdas, B. (2010). More on 'big things': Building events and feelings. *Transactions of the Institute of British Geographers, 35,* 334–349.

Runeson, G. (1997). The role of theory in construction management research: Comment. *Construction Management and Economics, 15*(3), 299–302.

Sage, D. (2013). 'Danger building site—Keep out!?': A critical agenda for geographical engagement with contemporary construction industries. *Social and Cultural Geography, 14*(2), 168–191.

Sage, D., Dainty, A., & Brookes, N. (2010). Who reads the project file? Exploring the power effects of knowledge tools in construction project management. *Construction Management and Economics, 28*(6), 629–639.

Sage, D., Dainty, A., & Brookes, N. (2011). How actor-network theories can help in understanding project complexities. *International Journal of Managing Projects in Business, 4*(2), 274–293.

Sage, D., Dainty, A., & Brookes, N. (2013). Thinking the ontological politics of managerial and critical performativities: An examination of project failure. *Scandinavian Journal of Management, 29*(3), 282–291.

Schweber, L. (2015). Putting theory to work: The use of theory in construction research. *Construction Management and Economics, 33*(10), 840–860.

Sennett, R. (2008). *The craftsman.* London: Allen Lane.

Seymour, D., & Rooke, J. (1995). The culture of the industry and the culture of research. *Construction Management and Economics, 13*(6), 511–523.

Seymour, D., Crook, D., & Rooke, J. (1997). The role of theory in construction management: A call for debate. *Construction Management and Economics, 15*(1), 117–119.

Seymour, D., Crook, D., & Rooke, J. (1998). The role of theory in construction management: Reply to Runeson. *Construction Management and Economics, 16*(1), 109–112.

Sherratt, F. (2015). Legitimizing public health control on sites? A critical discourse analysis of the Responsibility Deal Construction Pledge. *Construction Management and Economics, 33*(5–6), 444–452.

Silver, M. (1986). *Under construction: Work and alienation in the building trades.* Albany, NY: Suny Press.

Smiley, J.-P., Fernie, S., & Dainty, A. (2014). Understanding construction reform discourses. *Construction Management and Economics, 32*(7–8), 804–815.

Smith, N. (1986). Gentrification, the frontier, and the restructuring of urban space. In N. Smith & P. Williams (Eds.), *Gentrification of the city*. Boston: Allen and Unwin.

Smith, L. (2013). Trading in gender for women in trades: Embodying hegemonic masculinity, femininity and being a gender hotrod. *Construction Management and Economics, 31*(8), 861–873.

Strebel, I. (2011). The living building: Towards a geography of maintenance work. *Social and Cultural Geography, 12*(3), 243–262.

Styhre, A. (2009). Tacit knowledge in rock construction work: A study and a critique of the use of the term. *Construction Management and Economics, 27*(10), 995–1003.

Styhre, A. (2017). Thinking about materiality: The value of a construction management and engineering view. *Construction Management and Economics, 35*(1–2), 35–44.

Suchman, L. (2000). Organizing alignment: A case of bridge-building. *Organization, 7*(2), 311–327.

Sykes, A. (1969a). Navvies: Their social relations. *Sociology, 3,* 157–172.

Sykes, A. (1969b). Navvies: Their work attitudes. *Sociology, 3,* 21–35.

Thiel, D. (2007). Class in construction: London building workers, dirty work and physical cultures. *British Journal of Sociology, 58*(2), 227–251.

Thiel, D. (2010). Contacts and contracts: Economic embeddedness and ethnic stratification in London's construction market. *Ethnography, 11*(3), 443–471.

Thiel, D. (2013). *Builders: Class, gender and ethnicity in the construction industry*. London: Routledge.

Tönnies, F. (1963). *Community and society* (C. Loomis, Trans. & Ed.). New York, NY: Harper Torchbooks.

Tryggestad, K., & Georg, S. (2011). How objects shape logics in construction. *Culture and Organization, 17*(3), 181–197.

Tryggestad, K., Georg, S., & Hernes, T. (2010). Constructing buildings and design ambitions. *Construction Management and Economics, 28*(6), 695–705.

UCATT. (2017). Women in Construction, Union of Construction, Allied Trades and Technicians. Retrieved from https://www.ucatt.org.uk/women-construction.

UKSIC. (2017). UK Standard Industrial Classification of Economic Activities 2007, Office of National Statistics. Retrieved from https://www.ons.gov.uk/methodology/classificationsandstandards/ukstandardindustrialclassificationofeconomicactivities/uksic2007.

Vitruvius. (15 BCE/1960). *The ten books on architecture* (N. Morgan, Trans.). Toronto: General Publishing Company.

Waite, R. (2017). Number of women architects rises again. *The Architects Journal*. Retrieved from http://www.architectsjournal.co.uk/news/number-of-women-architects-in-aj100-practices-rises-again/10007191.article.

Walter, L., & Styhre, A. (2013). The role of organizational objects in construction projects: The case of the collapse and restoration of the Tjörn Bridge. *Construction Management and Economics, 31*(12), 1172–1185.

Watts, J. (2007). Porn, pride and pessimism: Experiences of women working in professional construction roles. *Work, Employment & Society, 21*(2), 299–316.

Watts, J. (2009). 'Allowed into a man's world' meanings of work–life balance: Perspectives of women civil engineers as 'minority' workers in construction. *Gender, Work and Organization, 16*(1), 37–57.

Weber, M. (1978). *Economy and society: Volume 2* (G. Roth & C. Wittich, Eds.). Berkeley, CA: University of California Press.

Whyte, W. (2006). How do buildings mean? Some issues of interpretation in the history of architecture. *History and Theory, 45*, 153–177.

Willis, P. (1977). *Learning to labour: How working class kids get working class jobs*. New York, NY: Columbia University Press.

Wright, T. (2016). *Gender and sexuality in male-dominated occupations: Women workers in construction and transport*. Basingstoke: Palgrave Macmillan.

Yarrow, T., & Jones, S. (2014). 'Stone is stone': Engagement and detachment in the craft of conservation masonry. *Journal of the Royal Anthropological Institute, 20*, 256–275.

Zhang, X. (2005). Critical success factors for public–private partnerships in infrastructure development. *Journal of Construction Engineering and Management, 131*(1), 3–14.

2

'This Building Is Never Complete': Studying Adaptations of a Library Building Over Time

Hiral Patel and Dylan Tutt

An excerpt of an interview with a member of library staff, 4th February 2015:

Jiva: … and again this summer we will be doing it all again, I suppose.

Hiral: To move ground floor books, I think?

Jiva: Well, over the next two years it seems we are moving the whole library. Because I am not sure how they are going to achieve what they need to achieve without moving stock … so all the ceilings have to come down again.

Hiral: There's a sticker in some of the books that says, 'Book presented on the completion of the Whiteknights Building Library in 1960s' (Fig. 2.1). This building never completes.

[*We laugh*].

Jiva: Yeah…yeah.

H. Patel (✉) · D. Tutt
School of the Built Environment, University of Reading, Reading, UK
e-mail: h.a.patel@reading.ac.uk

© The Author(s) 2018
D. J. Sage and C. Vitry (eds.), *Societies Under Construction*,
https://doi.org/10.1007/978-3-319-73996-0_2

Fig. 2.1 Bookplate commemorating the completion of the library building (*Source* A member of staff at University of Reading Library)

A Building Is Not a Fixed Object

The physicality of a building, which culminates in the completion of a building project, leads to an illusion of the completion of the building itself. The exterior of a building may reinforce this belief in its static nature. Moving beyond the concept of buildings as fixed physical objects, this chapter will draw on a rich empirical study of the adaptations and refurbishments of a fifty-year-old library building at the Whiteknights Campus, University of Reading. Even many decades after the library building was built, certain facades have not changed (Figs. 2.2 and 2.3).

Often, a newly constructed building seems like a static object, simply a product of the design process. Yaneva (2005) describes how an

Fig. 2.2 Library building, University of Reading, 1960s (*Source* University of Reading, Special Collections, MS 5305 (University History))

abstract building can be concretised, by focusing on the practice of scaling up and down using architectural models, during her ethnographic observations in an architectural office:

> The scaling venture is long lasting, but not infinite. Scales vary until they are 'stabilized' at a certain level of definition of the building. Then the architects stop scaling and 'fix' the building. (Yaneva 2005, p. 887)

However, when one looks at the life of a building, such *stabilization* need only be temporary, or a momentary pause in the scaling process. A newly completed physical building may be seen as just the beginning of that building. The fixity of a building is further problematised when adaptations of that building are brought forth.

Fig. 2.3 Library building, University of Reading, 2014 (*Source* Patel)

In their survey of 'terminal literacy' in architecture, Cairns and Jacobs (2014) explore the ways in which buildings decline. 'Dross', 'rust', 'subtraction', 'wasting', 'junk' and 'event' all point towards the incremental decline (and creation) of buildings. Presenting a shift away from the conception of buildings as static objects, Maudlin and Vellinga (2014) brought together studies of 'occupation', 'appropriation' and 'interpretation' of architecture. Their introduction articulated their intent clearly:

> Consuming Architecture seeks to step beyond the role of the architect altogether and understand how buildings are consumed by society as a whole (a society that includes architects but is not overly preoccupied with them or their internal professional concerns). (Maudlin and Vellinga 2014, p. 5)

Furthermore, they do not burden the notion of 'consumption' with negative connotations of destruction, decline and decay, but rather conceive of consumption as generative. In consuming architecture, a physical building interacts with discrete groups of people, such as inhabitants, builders, and critics, and in turn becomes imbued with new meanings and values.

The empirical work tracing the changes in the library building poses a theoretical challenge, namely to better account for and address the fluidity of a building, both spatially and temporally. To address this problem, an innovative methodology was developed to study instances of adaptations over time. This required the researcher to adopt multidisciplinary approaches and concerns, adopting and critiquing multiple positions of expertise and inquiry (ethnographic, historical, architectural, etc.) and working with rich and varied data sets. Ethnographic methods were used to explore the library's present practices, including 'shadowing' and interviewing users (with regard to browsing books, issuing books, using study desks) and staff (with regard to book moves, fire safety audits); attending library refurbishment project and university committee meetings; analysing artefacts, documentary sources and internet sources. To understand the library's past practices, these approaches were coupled with historical methods, which involved analysing organisational and design archives; interviewing previous library staff and university members; and analysing artefacts and the physical building itself as a historical data source. In this chapter, we will draw mainly from visual data in the form of images, both from the archives and from the ethnographic fieldwork, to help illustrate our arguments. One further mode of empirical engagement which we will be drawing on is that of curating an exhibition about the history of the library building, which was held on 11th November 2014 within the library building to mark its 50th anniversary—akin to 'exhibiting the library in the library'.

An ontology of the library develops in which different versions of the library are relationally and multiply enacted. In turn, this contributes to the wider social sciences through developing theoretical understandings of enactment and foregrounding the role of buildings as the locale of overlapping and contesting practices. Reconceptualising a building, in our case the library building, as unfinished and always in flux, provides a new avenue for theoretical understandings of adaptations of buildings and the implications for design practices for adaptability.

Studying Buildings Over Time: Key Research Concerns

This study focuses on two interrelated concerns pertaining to the built environment: linking design for adaptability of buildings to practices of adaptations of buildings, and post-occupancy evaluation of buildings over longer periods of time. While considerable research has been carried out to devise design strategies for adaptability in buildings (for extensive review on this topic, see Schmidt 2014), there are few empirical studies that examine how that designed adaptability gets used after the building is constructed. As Gorgolewski (2005) explains, empirically examining how buildings are adapted can provide a new knowledge base for designing for adaptability:

> Analysis of how buildings are used, how they function, and how users wish to change them can provide designers with an insight and respect for the effects of time, and their designs may become more durable and capable of being adapted for changing requirements. (Gorgolewski 2005, p. 2812)

Kincaid (2000) echoes the need for research into how buildings are used, especially due to the advent of new technologies for working, in order to inform the design of buildings. Kelly et al. (2011) also advocate the benefits of studying how a building adapts over time and, in addition, present a pilot study highlighting the differences between how a building was designed to adapt and *how it was actually adapted*. It, thus, becomes crucial to empirically investigate adaptations of buildings to inform building design practices.

Post-occupancy evaluations are an endeavour to link design with the performance of buildings after they are constructed. Schneider and

Till (2007) collated 74 international housing case studies in which flexibility design principles were adopted. However, they do not explore whether the designed flexibility was used as intended. They acknowledge that post-occupancy evaluation of flexible housing is very rare and was beyond their research scope (p. 9). Habraken (2008) argues that the key criterion for flexibility is the distribution of control amongst the designers and users, and that post-occupancy research and long-term user feedback can generate understanding of user practices and inform design for flexibility. Very few studies revisit buildings beyond five years of its construction. For example, in 1998 the Elizabeth Fry Building on the University of East Anglia campus was surveyed as part of the Post-Occupancy Review of Buildings Engineering (PROBE) project. The building, which was commissioned in 1995, gathered much attention during the late 1990s for achieving excellent energy performance and good comfort levels on Building Use Studies (BUS) survey metrics (Standeven et al. 1998). In 2011, the PROBE team re-evaluated the building's performance (Bordass and Leaman 2012). In 1998, the building housed lecture rooms, seminar rooms, offices, dining rooms and a kitchen. However, in the period between the PROBE studies, its spaces underwent substantive changes of use: the seminar rooms were converted to administration offices, and the kitchen and dining rooms were changed into open-plan office.

Overall, the building's occupancy had increased, which had resulted in a decrease in perceived occupant comfort. The building's former kitchen had three small windows, a number that seems inappropriate for administrative offices. In addition, the building's acoustics have been affected due to its increased occupancy, and the replacement light fixtures and additional radiators installed in the interim had higher power ratings than those recommended in the original building design. Following the 2011 PROBE reassessment, the authors posed a question:

> ... the question arises as to whether the building should have had a more uniform pattern of windows to facilitate changes. On the other hand, does management really need to alter buildings so much? (Bordass and Leaman 2012, p. 36)

In this case, building performance was considered in terms of energy consumption and occupant satisfaction. However, a building's energy performance depends not only on the physical building itself, but also on the various practices which a physical building becomes part of and party to after its construction. A qualitative study of UK-based architectural practitioners engaging with post-occupancy evaluation revealed that there was frustration amongst many participants that the current post-occupancy evaluation toolkits appeared to favour quantitative measures for energy use and occupant satisfaction (Hay et al. 2017). The study also reported that there was interest in developing post-occupancy evaluation methodologies that go beyond the technical aspects of the building and cover a broader understanding of how a building works for the users.

Olsen and Bonke (2011) studied Danish experimental housing buildings which were constructed in 1983 using an 'open system' approach for flexible design. Going against the rhetoric of design adaptability as an undisputed benefit, they sought to understand if the designed adaptability was used as intended for these buildings following their construction 25 years ago. To access the experiences of those involved in the building programme, they interviewed five of the building owners, a member of the 1994 evaluation team and a civil engineer who served on one of the competition teams. Their findings suggest that in some cases the designed flexibility was used as intended. However, other technical innovations in the structural system of these buildings were not used as planned. In addition, the researchers found limitations in the buildings' original designs. Their study is novel both in its attempt to evaluate the buildings twenty years after construction and in analysing the experiences of individuals who have been involved with the building in the past. But their account is limited as it does not discuss 'why' in-built flexibilities were used or not used, or 'how' certain flexibilities were used. More importantly, their work to study adaptations of those houses empirically sparks a new research agenda: the need to reconceptualise our understandings of building adaptability as residing solely in the physicality of a building and to explore the methodological challenges to pursuing research in this area.

A Methodology for the Study of Building Adaptations

Moving beyond the limited concept of buildings as static objects, that are finished on the day the construction project is completed, leads us towards acknowledging that users are going to actively live with the building and, in turn, the building is going to live with them. Latour and Yaneva (2008) problematise the static nature of buildings and make a case for conceptualising buildings as being in flux. Using the analogy of photography for studying the flight of gull, they posit the need for theoretical tools to study the transformations of buildings. Moreover, studying adaptations of a building, as something constantly changing, also poses a distinct methodological challenge. The two issues of method and theory are inextricably linked here, for method is involved in enacting the reality about which we theorise (Law 2004). This applies whether we are to foreground the realities of the weathering of building materials (Mostafavi and Leatherbarrow 1993), obsolescence (Thomsen and van der Flier 2011), energy performance and occupant satisfaction (Cohen et al. 2001), or architectural consumption (Maudlin and Vellinga 2014).

Having trained in the field of architecture, Patel first relied on her skills and expertise in order to 'read' two-dimensional building representations—i.e. plans, elevations and sections—to trace the adaptations of a building. Beyond this disciplinary literacy, Zeisel (1984) conceives a method for analysing building adaptations by observing their 'physical traces'. He mobilises the analytic concepts of 'props' (i.e. things added or removed by users), 'separations' (i.e. dividers of space introduced by users) and 'connections' (i.e. connectors of space introduced by users). However, Zeisel's (1984) method has two limitations with regard to pursuing the aims of this study. First, his framework is intended to study the physicality of a building and overlooks many of its other aspects. This is not to say that a building's physicality is unimportant, but a sole emphasis on physicality is problematic. The reason for this limitation can be ascribed to the assumed ontology of a building, i.e. what a building 'is', which for Zeisel (1984) is limited to the

Fig. 2.4 1980s extension (*Source* University of Reading, Special Collections, MS 5305 (University History))

physicality of that building. The second problem with Zeisel's method is that it seeks to observe 'what' changes, but does not address the questions of 'how' and 'why' those changes occur.

If the tracing of changes to the physicality of the library building was to be undertaken, the episodes such as the 1980s extension to the library building (Fig. 2.4) would become the key focus for theorising building adaptations. However, architectural training and practice had engendered in Patel a belief that building project teams 'create' physical buildings that influence their users, who in turn continue to alter or appropriate them. The physical-trace observations method thus becomes problematic and limiting for our inquiry. Our empirical observations, stemming from ethnographic and historical methods, highlighted a multitude of adaptations to the physical building of varying degrees of intensity and levels of intervention. Not least, in revealing and accounting for how the construction work and refurbishment of the physical building was undertaken in tandem with the work done by the users and the staff, so as to continue the practices of the library (see Figs. 2.5, 2.6, 2.7, and 2.8).

Fig. 2.5 1960s book moves from the old library at London Road Campus to Whiteknights Campus. Once the physical building was built, the library staff had to move the books on trolleys (*Source* University of Reading, Special Collections, MS 5305 (University History))

During the course of the empirical fieldwork, the researcher reflected on her shift in (multi)disciplinary thinking, or ways of seeing, to account for how the library building changes over time. A move from the conception of a building as a 'physical building' (whose existence is not contingent on the practices involving it) to a 'building made in practice' (where a building exists only within the practices), represents a stark transition. It required, following Orlikowski (2010), moving from an engagement with practice as a 'phenomenon' (i.e. one sensitised to practice, or how buildings relate to the practices occurring around them) to a view of practice as 'perspective' (i.e. as an analytic concept, or how buildings themselves shape practices and are shaped by these

Fig. 2.6 1960s book moves from the old library at London Road Campus to Whiteknights Campus. Once the physical building was built, the library staff had to move the books on trolleys (*Source* University of Reading, Special Collections, MS 5305 (University History))

same practices), and even of 'philosophy' (i.e. practices as a world view in which buildings exist within practices).

Enacting the Library: Praxiographic Inquiry and the Politics of Empirical Work

The empirical fieldwork richly reveals that the library building is not a single object. Rather, there are multiple versions of it: book stack, reading room, workplace, café and so on. These different versions of the library can interact, overlap and/or clash with each other. However, they

Fig. 2.7 Temporary sign in the library building, displayed during 2014 refurbishment of the fourth floor. The area of construction work is cordoned off for users requiring them to find study areas on other floors. Alternative provision was made for accessing out-of-bounds books (*Source* Patel)

Fig. 2.8 Graphics on a wall of the library building, 8th August 2017. Construction work was to be carried out while the students study in the library and they were ensured that the library will stay open during the works (*Source* Patel)

The library building is a book stack.
The library building is a reading room.
The library building is a workplace for staff.
The library building is a place to meet over coffee.
The library building is a statistic in the annual report.
The library building is a score in the National Student Survey.
The library building is a venue for exhibitions and ceremonies.
The library building is a landmark on the itinerary of open day tours.
The library building is a target for student protests and demonstrations.
The library building is heritage to be conserved during the refurbishment project.
The library building is square metres recorded in the University space management database.

Fig. 2.9 Multiple versions of the library building (*Source* Patel and Tutt)

are not mutually exclusive; they have a 'fractional coherence'. So that, following Law (2002, p. 3), the library building 'balances between plurality and singularity. It is more than one but less than many' (Fig. 2.9).

Mol's (2002) ethnographic work on the disease atherosclerosis helps to explain how multiple versions of a building can coexist, and this can suggest a novel way of conceptualising buildings. She discusses how different versions of atherosclerosis were enacted in different parts of a hospital: in a clinic, a patient describes the pain experienced, and in a pathology department, a cross-section of diseased artery is examined. Rather than approaching atherosclerosis as an object ready to be measured and studied from different perspectives, Mol (2002) employs, what she describes as, a praxiographic approach to understand how different versions of atherosclerosis are enacted in practice and how these versions can coexist:

> If practices are foregrounded there is no longer a single passive object in the middle, waiting to be seen from the point of view of seemingly endless series of perspectives. Instead, objects come into being – and disappear – with the practices in which they are manipulated. (Mol 2002, p. 5)

When applied to buildings, such an approach helps to widen the (narrow) focus on the physicality of buildings. Indeed, rather than seeing a physical building as an object ready to be studied from different perspectives, one can examine how the physicality of a building, in this case the library building, is involved in multiple enactments of the library and how these enactments relate to each other. Mol's (1999) phrase 'ontological politics' refers to co-ordination and clashes between different versions of an object and points us towards an ontological understanding of reality as enacted in practice. This is the political dimension of this inquiry: it does not privilege different perspectives on the library building, but rather examines different versions of the library (Mol and Mesman 1996). The question of 'what a building is for someone' shifts to 'what it becomes in practice'. For this research, we did not define what a library is from the outset, but rather located it in its practices (Mol 2002, p. 32). The library building emerges in the enactments, such as issuing books, using study tables and in curating exhibitions. If those enactments were to disappear, the library building would disappear. And in those enactments, the physicality of the building interacts with other entities. Similarly, for this research we also did not define other entities related to the library, but rather located them in their respective practices. In this way, the library building becomes heterogeneous, and this conception of the building allows it to expand to include its interactions with other entities.

The Library Multiple—Over Time

We will now empirically consider two different enactments of the library over time (from 1958 and 2013). In fitting with the non-linear adaptation of buildings in practice, they illustrate the reoccurring themes of competing practices and contested spaces of the library.

Excerpt from fieldnotes made while referring archives, 10th September 2014:

It's the 23rd October 1958, 11:40 a.m. The Library Steering Committee, planning for the new library building at Whiteknights Campus, is meeting in the Committee Room at London Road Campus. They consist of the Vice-Chancellor, the Registrar, the Bursar, the Librarian and three other senior academic members of the university. The Architect and Assistant Registrar are in attendance by invitation, making it a total of nine attendees. This is third meeting of the Committee. Previously, the Committee had discussed the requirements, sketches of two schemes: compact shelving and traditional shelving. They had then visited other university libraries to inspect the compact shelving installation. The Chairman reports on the Committee members' visit to other university libraries; he also discusses the advantages and disadvantages of the compact shelving. The discussion then moves towards the planning of the new building. The Vice-Chancellor starts annotating his copy of the 'Agenda' (Fig. 2.10). A discussion ensues as to whether the library is a bookstore or a reading room?

***

Excerpt from Minutes of the Library Steering Committee, 23rd October 1958. Source: Box 256, University Records Centre:

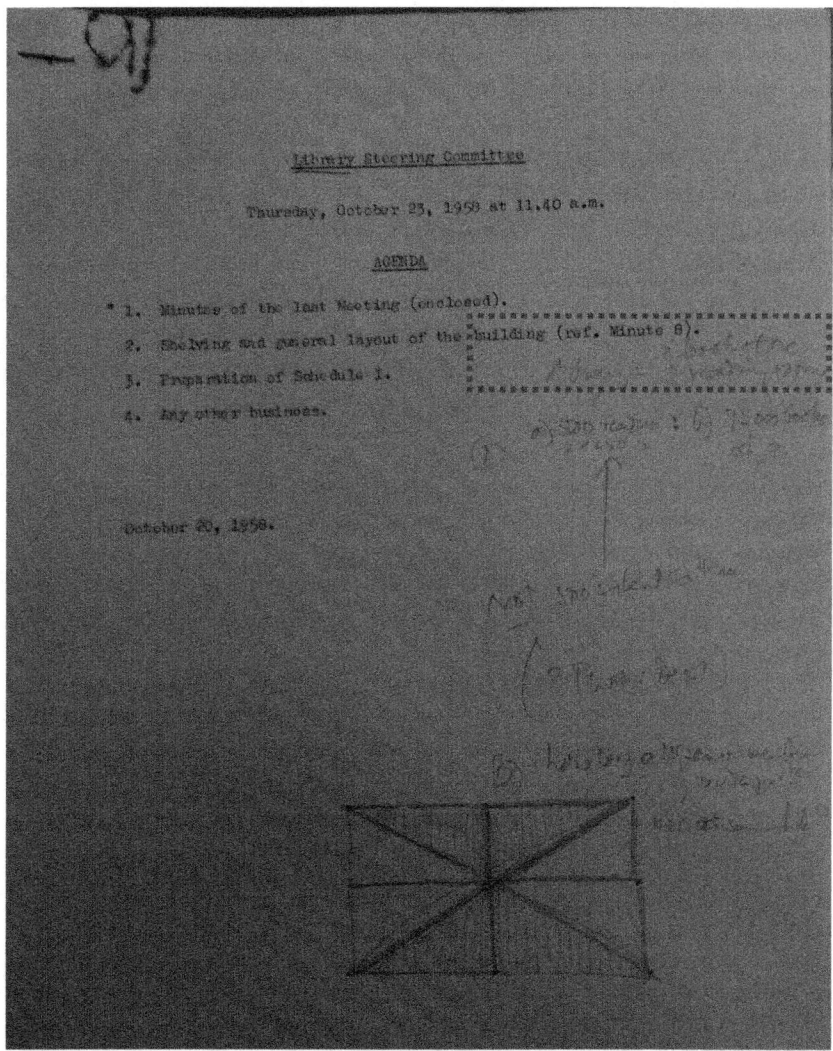

Fig. 2.10 Annotated library steering committee agenda 23rd October 1958. Highlighted annotation reads: 'library = ?bookstore; ?reading room' (*Source* Box 256, University Records Centre)

The question was then raised whether in the planning of the Library so far; the fundamental principles which should govern the design of the building had been sufficiently discussed and decided. The following points were made:-

a) It could probably not be said that the Library should be primarily a book store or primarily a reading space since it would need to combine both functions.

The two versions of the library seem to be very different. The plan is to have 'stacks' for 450,000 volumes, and these shelves of books are inaccessible to the reader. This then creates the notion of the bookstore. The reading room would have space for readers, who may freely use the space to browse and refer to the books on the shelves or to just do their own work. It is not necessary to use the books in the stack in order to use the reading room. However, these two versions of the library are not mutually exclusive and have to be combined in the library.

Excerpt from fieldnotes:

Nearly fifty-five years later, it's 22nd October 2013, 11:30 a.m. I sat in Committee Room 2 at Whiteknights House on the Whiteknights Campus. It is the final meeting of the pre-feasibility study for remodelling three university buildings, one of which is the library. Soon the whole room fills up and extra chairs have to be brought in. I did not count the total number, but more than fifteen attendees were present. One of the aims of the pre-feasibility study was to establish the future needs of the library for twenty-first-century staff and students. The study culminated in producing a document that included 'Statement of needs'. Similar to the discussions of 1958, this 'Statement of Needs' also focuses on the provision of study spaces. However, the concerns regarding the 'bookstore' have disappeared and instead 'e books' and 'e resources' are discussed.

In comparing these two temporally distant episodes, we can trace the adaptation of the library. The practices of using 'e books' involve entities which are different than those involved in using printed books and in turn enact the library differently. By her own admission, Mol's (2002) work on atherosclerosis does not look at how it changes over time, whether within the body or in the medical field:

In this book I do not go into the history of the diseases I describe. I even flatten out most of the changes observed over the few years of my fieldwork... But in this book the matrices produced are primarily spatial. (Mol 2002, p. 25)

The study of the library building is not just about multiple versions of the library at any given time, but also about the changes in these versions over time. In the library's case, the matrices Mol refers to are both spatial and temporal. An analogue here is de Laet and Mol's (2000) study of the Zimbabwe Bush Pump. The boundaries of the pump are not solid and sharp: it serves as a hydraulic pump, sanitation device, a nation-builder and a community pump, and the boundaries of each of these versions of the pump overlap. A library building has to be secure to prevent the theft of its books. To achieve this aim, the physical building, along with other entities, enacts a boundary separating 'inside' and 'outside'. In the case of Reading University Library, this boundary can be traversed only by moving through the exit panels at the main entrance. Other exit doors in the building are kept locked and cannot be used by the library users. Unlike the Bush Pump example, establishing new relations with entities is resisted in this enactment of the library.

However, like the pump, the library building is somewhat fluid. In the case of a fire alarm, the boundary to securing books is suspended. Doors which are otherwise locked are opened to evacuate the library users. Conceptualising the building as fluid affords an understanding of the adaptation of the building in such a way that continuity (as a secure bookstack most of the time) and change (as a safe building for users during the event of fire) are simultaneously addressed. In the enactment of the library when issuing books, the building is not fluid at all times; this fluidity is temporally controlled. It is fluid when the fire alarm rings.

While Mol's (2002) praxiographic study of atherosclerosis is based on her ethnographic observations, she indicates that historical methods are another way to explore reality through practice (p. 158). Indeed, Mak (2006) employs Mol's praxiographic approach to study a nineteenth-century medical case history of a hermaphrodite in order to examine different enactments of sex in the medical practice. By analysing this case history, Mak studies the practicalities and technicalities involved in deciding/doubting the 'true' sex of the patient and concludes that different medical practices enact different versions of the patient's sex. However, to pursue a praxiographic study of the library building over its history required a combination of ethnographic and historic methods, in order to empirically access and compare temporally distant practices.

Our decisions to research certain enactments of the library instead of others were made during a methodological process, working through the practicalities of, and developing innovation in, fieldwork. The position of the researcher as library user was leveraged in order to check out books and to use study tables in the library building, etc. However, exploring the enactment of the library through the practices of curating an exhibition represented a unique opportunity; and again, one where direct access to this enactment was available. Yet, the enactment of curating an exhibition only gained analytical significance during a period of reflection after it was over. Methodological adaptation and innovation were required in fieldwork again, to then trace the adaptations of the building during past exhibitions.

Building an Exhibition

Excerpt from voice diary, 5th November 2014:

We [a friend and I] discussed the setup [of the exhibition] in the actual space. My friend gave me some suggestions. She was concerned about the visibility of the exhibition for those coming and going. So we swapped the bookshelves to the catalogue side (see Figs. 2.11, 2.12, and 2.13). The foam boards, on the other hand, were lower, so that people could still see the shelves and to visually connect to what's happening in the exhibition area. So we made that swap.

To improve the visibility of the exhibition, to mark the 50th anniversary of the library building, the shelves with exhibits were placed in front of the library catalogues and in turn visually screened them. As a result, the movement of the library users in this space was altered. In curating

Fig. 2.11 Curating an exhibition in the library. Setting up the exhibition (*Source* Patel)

the exhibition, Patel was able to manipulate the physical building by changing the layout of the main hall on the ground floor. Thus, in making exhibitions, the exhibitors were able to adapt the physical building.

Excerpt from voice diary, 19th November 2014:

I think the exhibition brought with it a rush. It was quite physical. It was embodying [sic]. I have lost three to four kilos (of body weight) in one week … I am now going back to my reading and writing…I don't think it is so physical. …I have been at home for two days now. I don't see a need to go to office because actually I am doing the work that I would do in the office here which itself is quite different to the exhibition where I had to be in the bindery to do the work. There was this whole physicality in it, which has been lost.

Fig. 2.12 Curating an exhibition in the library. Setting up the exhibition (*Source* Patel)

Curating the exhibition required Patel to work in and with the physical building. This empirical engagement with the physicality of the building was different than other modes of data collection such as studying archives or interviewing members of the library. In addition

Fig. 2.13 Exhibition open (*Source* Annual Review 2014–2015, University of Reading Library)

to leading to theorisation, empirical engagement of this kind is akin to construction work: the labour (putting up shelves, mounting/moving exhibits), the site (the ground floor of the library), the assembly of materials and equipment (models, table, chair, microfiche reader, boards, books, cards, drawings and computer). In other practices of the library, the extent to which users can adapt the building may be limited or even prohibited. When issuing a book, the users are prohibited to make any adaptations in order to prevent theft of books. In order to issue a book, the users need to scan their ID cards and barcodes of books at the self-issue terminals and exit the building through electro-magnetic panels which are continuously monitored by the library staff. However, through the practices of using and occupying the tables for studying, the users can temporarily adapt the physicality of the build-ing through 'marking' the territory they claim (Goffman 1971).

They can move chairs, converse at the tables designed for individual study, and leave their personal items on the table while they are away to park the table. But such territory may be limited to the table or the study furniture or other pieces of the 'stall' to which they can adapt and lay temporary claim. It was only through the enactment of the library as an exhibition venue that selects users could adapt the physicality of the building to a much greater extent. In our research, we conceptualise the changes of a building over time, to offer fresh insights on our active role in the adaptability of buildings and how different versions of a building are enacted in practice. Yet, this example of curating an exhibition in the library, in a built environment research context, also offers fresh methodological insights on the practice of ethnographic methods. In theorising the body as a tool of inquiry, this particular empirical work is revealing not just of the embodiment of the everyday work in the library, but of the affordances and materiality of the building and the embodied knowledge of adapting a building in practice.

Exhibiting the Library in the Library

Presenting an exhibition about the library in the library building allowed artefacts from various moments in the life of the building to be juxtaposed against each other: the paper ID cards from 1960s were displayed next to the recent campus ID cards; to depict involvement of library in student campaigns, photos of graffiti from 1970s were displayed next to a video of a flash mob from 2011; unrealised design options were displayed in the building which was actually constructed. Artefacts, which are involved in multiple practices, also revealed their conflicted presence in their use across different academic disciplines.

Excerpt from personal voice diary, 12th November 2014:

[A microfiche reader was displayed in the exhibition] One visitor said that he couldn't believe that a microfiche reader had become an object in a museum, because they remain in constant use in his department, where they have monographs with microfiches. And I said that it might be discipline-dependent in that sense, because I had not seen a microfiche reader before encountering it in the field.... So, perhaps in his discipline this is quite relevant. But in my discipline and others, they aren't current.

Patel included a microfiche reader in the exhibition to demonstrate how microfiche catalogues were used during the 1980s, before they were replaced by computerised catalogues. Having encountered it in her fieldwork for the first time, its association for her was with a particular enactment of the library in the past practices of searching catalogues. However, in this instance, the visitor maintained that microfiche readers are still very much in active use for reading microfiches inside books and are part of his disciplinary practices. The microfiche reader highlighted the temporal specificity of the fieldwork (Rendell 2009) and the thorny nature of curating (different versions of) the library across time. This moment and the contention around this object also, perhaps, speaks of 'ontological interference' (Law 2004) and the multiplicity of objects and claims. As the different enactments of the library play out and simultaneously exist, the use and interpretation of the objects (e.g. microfiche reader) are never fixed, but are in interference with one another.

Curating the exhibition in the library offers a site for discussion, contradiction and clarification regarding the research process. Indeed, following Lees (1997), the library can be conceptualised as a heterotopia, as 'a space of simultaneity, phasing different spaces and times together' (p. 327). Books on library shelves are published across different periods of time. New tables, old tables. New glass screens and old partition walls. New carpet, old slab. New paint, old marble cladding. Moreover, the library as a venue for exhibition offered a site which was open to visitors from different academic disciplines. Similarly, library users can access books from various academic disciplines on the shelves of the library. As Lees (1997) observes at Vancouver Public Library, 'in many ways the intellectual space of the public library has the greatest potential to be a democratic public space, for it is relatively free, open and offers possibilities for contestation' (p. 341).

As with the case of microfiche reader, the changes in practices and use of materials and artefacts do not take a linear route, but rather adapt to the changing enactments of the library. The malleability of physical materials and architectural space was also revealed to be an important part of how a building changes over time. Again, with the example of library tables, the fieldwork reveals how adaptation does not necessary follow a linear path.

Refurbished Tables: Outliving Bricks and Mortar

Brand's (1997) 'layer' model has been influential in informing research on adaptations of buildings and for devising design strategies for adaptability. Brand conceptualises buildings as being comprised of six layers: site, structure, skin, services, space plan and stuff. The basis for these divisions is the relative rate of change across these layers. The former layers are more permanent, and the latter less so. For example, while the 'stuff' layer may change daily or monthly, the 'services' layer typically becomes obsolete every 7–15 years (Brand 1997). A table might be replaced in five to seven years (Duffy 1990), as it is part of the scenery layer. However, some of the library tables have been used for over fifty years (Figs. 2.14, 2.15, and 2.16). Here, the tables defy the deterministic approach towards the life span of a building proposed by Brand's (1997) layers model of the building. When the practices of enacting a study space are examined, the tables (layer—stuff), sockets (layer—services) and columns (layer—structure) are tethered to each other in a non-hierarchical manner. Moreover, the 1960s tables were recently refurbished, and as a result of adaptations made to the physicality of these tables (e.g. new surfaces, holes for power sockets and cables, dividers and lamps), they enable new possibilities and ways of using them. By making these adaptations, the enactment of the library in using these tables has gained a renewed life. The tables have remained but adapted, as many of the architectural aspects around them have undergone significant change, including a major building extension during the 1980s and new electrical wiring and new flooring during 2011–2015.

The material relationship between table and building is also intriguing. It is useful here to reconsider the meaning of *building* and *dwelling*

Fig. 2.14 Reading gallery in the 1960s (*Source* University of Reading, Special Collections, MS 5305 (University History))

as words and activities, as developed in Heidegger's influential essay in the area of architectural phenomenology, *Building, Dwelling, Thinking* (1971). As Scharr (2007) observes, through this lens we see objects being 'built according to the needs of dwelling and dwelt according to configurations of building' (p. 69). This leads Scharr (2007) to question, through his mobilisation of Heidegger, 'when a table might or might not be considered as architecture' (p. 41). This distinction between building and dwelling, or indeed building and table, would be deemed as unnecessary by Heidegger himself, who describes both as 'built things'. Yet, as Scharr (2007) explains, and our research of the library espouses, both are similar because 'they relate to people" in everyday life and should be understood 'through tactile and imaginative experience; not as a detached object' (p. 46).

Fig. 2.15 Reading gallery before the 2014 refurbishment (*Source* Patel)

Indeed, in his early writings, *Ontologies: The Hermeneutics of Facticity*, Heidegger (1988, p. 90) specifically examines the social, material and

Fig. 2.16 Reading gallery after the refurbishment (*Source* Patel)

spatial practices involving a table. In this case, it was the family table where places at the table were routinely occupied and connected to everyday activities and he recalls, from memories and visible marks, instances of eating, writing, talking and celebrating at the table. Yet, the spatial practices and socialities of buildings and objects are too often and too easily separated from aspects of physicality and materiality. Even if we consider the very word origin of 'refurbish', it carries a meaning of 'making do' and surface-level adaptation, traced back through the Germanic origin *furban*: 'to appear'. However, as Scharr (2007) observes, a building should not be understood as 'just as an object *to be admired* or the product of a construction management process' (p. 46). Equally, the refurbishment of the tables is not merely aesthetic and functional. Each 'built thing', each table, each building, enables certain activities to 'take place'. To return to our pyramid of versions of the library (Fig. 2.9), buildings create places to study, to remember, to protest, to entertain and to do business. The buildings in turn are enacted in these practices. Buildings not only take up space, they create space, open it up for human tasks and dwelling, and are engaged in everyday interactions and practices. It is not surprising that with the changing functions of the library in society, the library building is constantly changing as well.

Conclusion

A building as an enactment links the past practices with the present and into the future. Moreover, the adaptations of a building need not be linear. Rather, changes in a building can be nonlinear such that spaces revert back to the earlier uses. For instance, several exhibitions were held in the halls on second and fourth floor of the library until the 1970s. The glass screens separated these halls from the book stack while maintaining the visual connection (see Figs. 2.17, 2.18, and 2.19). Over time the halls became encroached with books and tables, and exhibitions were no longer held there. The glass screens were replaced by a solid partition wall during the 1980s. More recently, the solid partition wall has been removed and the glass screens have been reinstated. The books and tables are removed and, as a result, making those halls a potential venue for exhibitions once again.

The two empirical examples of curating the exhibition and refurbishing tables demonstrate the sort of insights and challenges that stem from conceptualising a building in flux. The praxiographic inquiry suggests that theory and practice are inseparable, as entities gain ontological emergence only in practice. The purpose of theory is not to tame the

Fig. 2.17 Fourth floor exhibition hall in 1960s. The glass screens, shown in the photograph, were replaced by opaque partition wall during the 1980s (*Source* University of Reading, Special Collections, MS 5305 (University History))

messiness of practices in which a building is enacted, but rather to gain from such messiness and make sense of them. The empirical engagement with practices plays a vital role in the theorisation of building adaptations. The conceptualisation of the university's library building, as enacted in practice, underscores how the adaptation of this building can be achieved through manipulating entities with/without changing the physical building itself. Such conceptualisation implies a need for renewed design practices for adaptability of buildings, as it challenges the efficacy of current adaptable design strategies which focus solely on the physical building. Within the research on designing for adaptability we would argue that, ironically, a lot of theoretical approaches simply do not account for the nuances of contestations between multiple versions of a building and multiple temporal trajectories as encountered in the practices of adaptation.

Fig. 2.18 In 2013, the partition wall was removed and the glass screens were installed (*Source* Patel)

We have shown how the location of practices, in a specific built environment context, is important. Yet the different versions of the library, the building/s, the objects, are relationally and multiply enacted. Empirical snapshots of exhibitions, glass screens, library tables, etc., illustrate the need for us to question (temporally) linear and normative assumptions on the use and adaptation of materials, objects and buildings in wider social science research. Through this study, we push for empirical visions and theoretical frameworks of research that look beyond notions of the fixity of the buildings and infrastructure, towards acknowledging their flexibility and heterogeneous nature over time, 'building' as always in the making. Indeed, any studies of practice and ways of living need to account for our interaction with, and active shaping of, the built environment and the multiple enactments in which they are entailed.

Fig. 2.19 A mock display was made in the exhibition hall for consultation with a few library staff, as part of the process of curating the 50th anniversary exhibition (*Source* Patel)

Acknowledgement This research was funded by EPSRC Doctoral Training Grant Studentship, award reference 1168162.

References

Bordass, B., & Leaman, A. (2012, March). Test of time. *Chartered Institute of Building Services Engineers Journal*, 30–36.

Brand, S. (1997). *How buildings learn: What happens after they're built*. London: Phoenix Illustrated.

Cairns, S., & Jacobs, J. M. (2014). *Buildings must die*. Cambridge, MA: The MIT Press.

Cohen, R., Standeven, M., Bordass, B., & Leaman, A. (2001). Assessing building performance in use 1: The probe process. *Building Research & Information, 29*(2), 85–102.

De Laet, M., & Mol, A. (2000). The Zimbabwe bush pump: Mechanics of a fluid technology. *Social Studies of Science, 30*(2), 225–263.

Duffy, F. (1990). Measuring building performance. *Facilities, 8*(5), 17–20.

Goffman, E. (1971). *Relations in public: Microstudies of the public order.* London: Allen Lane.

Gorgolewski, M. (2005). Understanding how buildings evolve. In *The 2005 World Sustainable Building Conference, Tokyo (SB05Tokyo)* (pp. 2811–2818), Tokyo, 27–29 September.

Habraken, J. (2008). Design for flexibility. *Building Research & Information, 36*(3), 290–296.

Hay, R., Samuel, F., Watson, K. J., & Bradbury, S. (2017). Post-occupancy evaluation in architecture: Experiences and perspectives from UK practice. *Building Research & Information,* 1–13. https://doi.org/10.1080/09613218.2017.1314692.

Heidegger, M. (1971). *Poetry, language, thought.* New York: HarperCollins.

Heidegger, M. (1988). *Ontology: The hermeneutics of facticity* (J. van Buren, Trans.). Bloomington: Indiana University Press.

Kelly, G., Schmidt, R., Dainty, A., & Story, V. (2011). Improving the design of adaptable buildings though effective feedback in use. In *CIB Management and Innovation for a Sustainable Built Environment Conference,* Amsterdam, The Netherlands, 20–23 June.

Kincaid, D. (2000). Adaptability potentials for buildings and infrastructure in sustainable cities. *Facilities, 18*(3), 155–161.

Latour, B., & Yaneva, A. (2008). Give me a gun and I will make all the buildings move: An ANT's view on architecture. In R. Geiser (Ed.), *Explorations in architecture: Teaching, design, research* (pp. 80–89). Basel: Birkhäuser.

Law, J. (2002). *Aircraft stories: Decentering the object in technoscience.* Durham, NC: Duke University Press.

Law, J. (2004). *After method: Mess in social science research.* Abingdon: Routledge.

Lees, L. (1997). Ageographia, hetereotopia and Vancouver's new public library. *Environment and Planning D: Society and Space, 15*(3), 321–347.

Mak, G. (2006). Doubting sex from within: A praxiographic approach to a late nineteenth-century case of hermaphroditism. *Gender and History, 18*(2), 332–356.

Maudlin, D., & Vellinga, M. (Eds.). (2014). *Consuming architecture: On the occupation, appropriation and interpretation of buildings.* London: Routledge.

Mol, A. (1999). Ontological politics. A word and some questions. *The Sociological Review, 47*(S1), 74–89.

Mol, A. (2002). *The body multiple: Ontology in medical practice*. Durham: Duke University Press.

Mol, A., & Mesman, J. (1996). Neonatal food and the politics of theory: Some questions of method. *Social Studies of Science, 26*(2), 419–444.

Mostafavi, M., & Leatherbarrow, D. (1993). *On weathering: The life of buildings in time*. Cambridge, MA: The MIT Press.

Olsen, I. S. & Bonke, S. (2011). Learning on flexibility from experiences—revisiting housing estates after 25 years. In *Architecture in the Fourth Dimension. Methods + Practices for a Sustainable Building Stock: Proceedings of the Joint Conference of CIB W104 and CIB W110* (pp. 36–40), 15–17 November.

Orlikowski, W. J. (2010). Practice in research: Phenomenon, perspective and philosophy. In D. Golsorkhi, L. Rouleau, D. Seidl, & E. Vaara (Eds.), *Cambridge handbook of strategy as practice* (pp. 23–33). Cambridge: Cambridge University Press.

Rendell, J. (2009). Constellations (or the reassertion of time into critical spatial practice). In C. Doherty & D. Cross (Eds.), *One day sculpture* (pp. 19–21). Bielefeld, Germany: Kerber Verlag.

Scharr, A. (2007). *Heidegger for architects*. London: Routledge.

Schmidt, R. (2014). *Designing for adaptability in architecture* (Ph.D. thesis, Loughborough University, UK).

Schneider, T., & Till, J. (2007). *Flexible housing*. Oxford: Architectural Press.

Standeven, M., Cohen, R., Bordass, B., & Leaman, A. (1998, April). PROBE 14: Elizabeth fry building. *Building Services Journal*, 20–25.

Thomsen, A., & van der Flier, K. (2011). Understanding obsolescence: A conceptual model for buildings. *Building Research & Information, 39*(4), 352–362.

Yaneva, A. (2005). Scaling up and down: Extraction trials in architectural design. *Social Studies of Science, 35*(6), 867–894.

Zeisel, J. (1984). *Inquiry by design*. Cambridge: Cambridge University Press.

3

Constructing Work: Politics, Society and Architectural History on the Paris Building Site

Jacob Paskins

In 1964, visitors to the *Arc de Triomphe* in Paris would have been distracted from their sightseeing by plumes of dust from deep excavations, diesel fumes from mechanical diggers and the jarring pulsations of pneumatic drills. Tourists' photographs would have been populated with the cranes, hoardings and concrete plants that surrounded the nineteenth-century monument, providing evidence of the construction of an underground regional express network—the future *réseau express régional* (RER). The celebrated structure in Place de l'Étoile that recalled consequential episodes in France's history was now witnessing the emergence of Paris's future. Construction of the RER was a spectacular sight both on and below the streets of the French capital, and it was by no means an isolated case. Similar scenes of construction were evident throughout the 1960s as major urbanisation projects progressed across the Paris region. In terms of the number and scale of building sites, Paris had not seen so much disruption and physical change

J. Paskins (✉)
University College London, London, UK
e-mail: Jacob.paskins@ucl.ac.uk

© The Author(s) 2018   **87**
D. J. Sage and C. Vitry (eds.), *Societies Under Construction*,
https://doi.org/10.1007/978-3-319-73996-0_3

since Haussmann's transformation of the city a century earlier during the Second Empire. Although the connection was slightly forced, some commentators certainly made a comparison between the nineteenth-century urban development of Paris and the urban master planning of the Paris region in the 1960s, led as it was by Paul Delouvrier, chief executive of the *District de la région de Paris*, the administrative organisation charged with overseeing much of the state-driven construction work. Robert Franc described Delouvrier as "the Haussmann of the suburbs" (1971, p. 165), while Michel Ragon wryly named the *boulevard périphérique*—the 36-kilometre ring road constructed between 1957 and 1973 that separated the city of Paris from its surrounding suburbs—"boulevard Delouvrier", echoing the more centrally located boulevard Haussmann (1968, p. 153).[1]

Across the Paris metropolitan region, a new, modern city emerged from the apparent chaos of building sites, in the form of new roads, bridges and underground car parks. Mainline railway stations such as Gare Montparnasse were rebuilt to fulfil increased transport capacity, and plans for a new airport at Roissy-en-France proceeded to serve the rapidly growing aviation industry. New power stations, factories and office buildings flourished, emulating the vast new high-rise business district at La Défense. Construction of new educational, health and leisure facilities was evidence of state modernisation plans responding to a rapidly expanding urban economy and population. Above all, the predominant aspect of urban transformation in the Paris region was the construction of housing, as private and public projects produced tens of thousands of new homes each year, many of which were located in large suburban housing estates, or *grands ensembles* (Fig. 3.1).

Visitors who saw the *Arc de Triomphe* held hostage by trenches and machinery may have realised that France had modernising ambitions, and that construction was a pillar of the French economy. Tourists, however, were probably less aware of the organisational difficulties of the construction industry. While passersby could not have missed the body of workers engaged with building the RER, onlookers might not as

---

[1]All translations are by the author.

Fig. 3.1 Construction of the *Tour Montparnasse* and shopping centre, Paris, 12 March 1971 (*Source* Coll. Pavillon de l'Arsenal, cliché DUVP)

easily have grasped the ways in which labour was complexly interwoven with the continual production of the city, at each level of the technical, political and social. Construction work in France in the 1960s raised questions about suburban growth, the industrialisation of building and about typological challenges for the provision of mass housing. But construction also revealed social and political concerns that were not limited to the architectural and building trades. Visiting the building sites of the Paris region in the 1960s, drawing on archival material, including newspaper reports, television programmes and trade union documentation, this chapter examines how construction work lay at the centre of debates about national identity, labour immigration and workers' rights. First, it analyses how a booming construction industry became a source of national pride in France, as political speeches and television reports about building sites fuelled the Gaullist rhetoric of national grandeur and progress. Building activity represented France's ambitions for national productivity, which translated into the development of construction techniques that the industry promoted to international markets. Many representations produced by the mainstream media and by the construction industry rendered workers invisible, despite the fact that workers were prominent on building sites and their lives were deeply implicated in the construction work they performed. Second, therefore, this chapter examines the human side of construction, asking how the Paris building site challenged representations that were imposed on it. What were the material realities of construction work, including employment practices, skills and training? The third section asks how workers organised themselves in the attempt to improve working conditions and to engage in international politics. Some activities on the building site expanded the duties of builders beyond daily labouring to include trade union representation and political campaigning.

While building sites offer insight into both construction processes and social structures, the temporary nature of construction means the traces of human activity can quickly become invisible and forgotten on completion of a project. If trade publications during the 1960s gave little acknowledgement to the role of the building worker, architectural and urban historians have also tended to overlook the organisation of construction and its related difficulties. The architectural humanities have only recently begun to acknowledge the role labour plays in the production of architecture and the city. Fourth, then, as a coda, this chapter addresses the challenges of writing an architectural and urban history of Paris through the lens of the building site. This chapter follows growing critical reflection on the question of "work" in architecture, confronting the often distant relationship between the activities of architectural design and the realisation of projects on the building site (Ockman 2015; Lloyd Thomas et al. 2016). Histories of the built environment must free themselves from the narrow constraints of architectural discourse to reflect on how architecture, urbanism and construction always interact with an array of social and political domains—and above all are deeply centred on the work of the building site.

A Booming Industry

In France, construction played a significant role in national economic productivity in the 1960s, enjoying 10% annual growth during the first half of the decade and contributing 8.7% of GDP (G. M. 1965). In 1962, building industries directly employed nearly 290,000 people in the Paris region alone (DGDP 1965, pp. 204–205); by the end of the decade, around 10% of the working population in France were employed in construction ("La main-d'oeuvre dans le bâtiment" 1974). Charles de Gaulle celebrated the visible results of the construction boom and the modernisation it brought to France; keen to show his government was taking a clear lead in urban development, the president made personal visits to building sites in the Paris region and announced in televised speeches to the nation statistics about national motorway construction and factory openings (*PL* 1966; "Le général de Gaulle visite le Paris de l'an 2000" 1966). The scale of construction output was principally evident in the annual statistics of completed housing units, which increased from 270,000 in 1957 to 500,000 in 1970 (RICS 1975, p. 158).

France was certainly not the only country to experience a postwar construction boom: considerable reconstruction and urban expansion took place in other European countries (Diefendorf 1993; Bullock 2002; Pilat 2014), the Soviet Union (Ritter 2012) and Japan

(Hein et al. 2003). In France, however, immediate post-war reconstruction focused far less on the Paris region than on the bomb-ravaged industrial cities and ports of northern and western France, such as Le Havre (Liotard 2007), Amiens and Nantes (Voldman 1997). Attention to the Paris region that suffered less widespread war damage was deferred to the late 1950s. When construction really took off in the 1960s, there was some political anxiety that France was someway behind the building progress achieved by neighbouring countries including Germany and Britain, and that France needed to compete with growing competition from international contractors and importers of foreign construction machinery and equipment.

At first glance, observers might well assume that the impressive construction rates in France were the result of modernised building trades that had embraced industrialisation to facilitate high output. Indeed, the French construction industry was undertaking significant research and development, marketing a number of proprietary construction systems—such as the widely used and exported Camus prefabricated panel process ("Procédé Camus" 1962)—and investing heavily in concrete production, making France the global leader of concrete manufacture in the mid-1960s ("L'industrie du ciment" 1966). Construction machinery, cranes on rails and innovative techniques in highway and tunnelling design were lavish calling cards for a modernising industry that was not afraid of publicising its achievements to an international audience through promotional articles and industrial films.[2] Within the national industry, impressive showcases such as the construction of the RER or *Tour Montparnasse* were nevertheless the exception rather than the rule as most construction companies at the start of the 1960s resembled traditional craft-based firms that had changed little in generations. The overwhelming majority of constructors in 1968 were still small artisanal outfits employing fewer than five workers; just 264 of the 264,000 construction companies in France employed more than

---

[2]For films celebrating the achievements of the French construction industry, see, e.g., Allain (1966), Chateau (1967), Leridon (1969). For a discussion of the use and interpretation of French moving image archives, see Paskins (2010).

500 people ("Enquête annuelle d'entreprise" 1973). These large general contractors gradually introduced mechanisation and the use of prefabricated components to the French building site, but the majority of constructors retained conventional building practices. In 1965, industrialised construction techniques such as proprietary systems—elements prefabricated in off-site factories and assembled on site—accounted for just 15,000 housing units, or 4% of total annual housing construction in France, a figure that was significantly lower than the 10% use of prefabrication in the UK, 15% in the USA and 80% in the Moscow region (Vernholès 1965). While some French constructors hoped building production would one day become as streamlined as a mechanised automobile plant, the present state of the industry was clear: high output was not the result of industrialisation but was driven by a massive manual workforce.[3]

There was nothing miraculous about the scale and pace of construction work and urban transformation in the Paris region; it depended on sheer manpower. As building sites opened across city and suburb, urban change became part of everyday life, forming the backdrop to neighbourhoods, becoming the focus of newspaper and television reports and taking prominent roles in feature films. Yet, despite this abundant new scrutiny of construction, representations in both the popular and professional media tended to ignore, even eliminate, the prominent evidence of human activity on building sites. Television cameras were frequent visitors to Parisian building sites, producing programmes that made a growing national television audience aware of the considerable physical changes to the city. Television was predominantly interested in spectacular engineering challenges, such as motorway junctions and flyovers ("Rétrospective Paris-chantiers" 1966); vast sports stadia ("Trois mille tonnes en mouvement" 1966); or the unprecedented development of the national wholesale food market at Rungis ("Les nouvelles halles de Rungis" 1966). These were one-off undertakings that demonstrated unusual or previously untested

---

[3]For debates about the need to increase the industrialisation of construction, see *JO* (1963) and Gall (1965).

construction processes whose complexity and scale no doubt made good television; but these spectacles—what David Nye (1994) calls the "technological sublime"—were once more not representative of most building sites. Even if television crews visited less high-profile sites, such as a new housing development, the shots and narrative of the edited programme would more likely focus on the architects' plans or site visits by dignitaries than on specific activities of workers on the building site ("Malraux à Pontoise" 1967). A typical television news report about the construction of a bridge would feature an interview with an engineer—wearing a clean suit, smoking a pipe with one hand, the other placed firmly in a pocket—interspersed with shots of construction procedures that emphasised machinery and structural components, with a commentator reeling off unfathomable statistics about the weight and size of structures ("Pose dernier élément pont Charenton" 1968). The televisual *mise-en-scène* emphasised the apparently "immaterial" work of the engineer (Deamer 2015) and provided little indication of the physical labour of builders working on the project. Much television construction footage resembled photographic reports of site visits in the architectural and construction trade press that described technical procedures in a scientific manner in an effort to promote the productivity of a modernising industry. Trade publications effectively reduced any workers that might be visible in photographs to an incidental role that merely facilitated a mechanised process calculated in advance by engineers and designers. Eliminating traces of manual labour, such representations of construction asserted confidence in modern processes and celebrated the "triumph" of French technical expertise.[4] Carefully choreographed views of the building site provided both professional and non-professional audiences with a myopic impression of construction that made workers invisible (Fig. 3.2).

---

[4]The staging of construction workers has a long tradition in photography. For an analysis of how civil engineers in nineteenth-century France used photographs of construction as a means of managerial control, see Weiss (2015).

Fig. 3.2 Men in suits observe workers on one of the many construction sites of the boulevard périphérique, 18 May 1967. The structure would form the approach road to the downstream bridge over the Seine between quai du Point du Jour and quai d'Issy-les-Moulineaux (*Source* Coll. Pavillon de l'Arsenal, cliché DUVP)

Intent on promoting modern French construction to international markets, industry leaders had little time to consider questions about the plight of their workforce. Trade publications rarely acknowledged problems in the profession related to labour and employment, perhaps for fear that they might detract from the positive narrative of the pursuit of "progress". Dazzled by the spectacular sights of the birth of a new city, spectators might easily have overlooked the prominent place workers held on building sites. Who were these workers, where did they come from, how did they live? How were the lives—and sociopolitical identities—of these workers implicated in construction work and the city they were building? Asking these questions starts to critique the predominant representations imposed on the building site by the industry and the French media, and offers a glimpse into the everyday realities of construction.

The Human Aspects of Construction

On the building site, a head count said it all. The French construction industry in the 1960s had a relentless appetite for large quantities of low-cost labour. Construction firms of all sizes, and especially the large general contractors, could draw on an abundant pool of workers to fuel the expanding industry. Migrant workers—skilled, or unskilled, with experience or with none—formed the backbone of French construction. A typical building site would be alive with languages from across Europe and Africa. The Parisian construction site was far more complex than the versions proposed by political cheerleaders and media representations as showcases for French modernity.

The material, social and political circumstances of the building site in and around Paris centred on migrant labour. Labour migration in France increased significantly in the three decades following the 1939–1945 war, rising from around 50,000 worker arrivals in 1950 to nearly 400,000 annual entries in 1965. Italian workers formed the first wave of post-war immigration, with over half a million Italian nationals moving to France between 1946 and 1965. After years of unauthorised immigration, Spanish and Portuguese workers accounted for two-thirds of official migration statistics during the first half of the 1960s. In addition to growing labour migration from south-eastern Europe, an estimated 50,000 workers from the former French colonies in sub-Saharan Africa arrived in France at this time. The total number of migrant workers in France is unknown as official statistics did not record the arrival of workers who were not registered with the *Office national d'immigration* (ONI, or National Immigration Department), or include workers who were from countries that did not benefit from officially agreed immigration quotas with the French government (McDonald 1969; *Le Monde* 1965b). Assessing the number of Algerian workers in France is similarly difficult as there was free movement between Algeria and France before 1962, but the years between 1946 and 1968 saw a significant increase in Algerians working in France, with an estimated 100,000 Algerians working in construction in 1965—nearly half of all Algerians officially recorded in employment, and a considerable proportion of the estimated half a million Algerians resident in France (*Le Monde* 1965a).

Construction was the most common source of employment for migrant workers as the expanding building trades relied on a large manual workforce. The director of the ONI claimed in a 1964 interview that French people were no longer willing to take "hard, dirty" and poorly paid work ("Ils sont trois millions" 1964). Some migrant construction workers sadly agreed with this view, tolerating often difficult living and working conditions in exchange for the promise of regular employment.[5] Around half of all migrant workers registered with the ONI were employed in construction (McDonald 1969, p. 122), and by the end of the 1960s, 38% of building workers in the Paris region were non-French nationals (FPB 1966, p. 41). Labour migration was a fundamental aspect of the Parisian building site, but accounts of the specific consequences immigration had on construction varied significantly. Education and skill levels of migrant workers were generally low, especially within the construction industry in the Paris region, where, according to the *Fédération parisienne du bâtiment* (FPB, the regional branch of the construction employers' organisation), 80% of migrant workers were classified as unskilled labourers, lacking basic literacy and qualifications (FPB 1966, p. 42). Low skill levels were nevertheless no barrier to work in construction, an industry which welcomed the plentiful availability of low-cost workers, and which was used to employing men with little education: a survey conducted by a construction workers' pension fund in 1966 found that only 1.6% of retired builders possessed technical qualifications (*Le Monde* 1967). The FPB expressed concern about the generally low quality of the workforce and blamed the extremely poor work safety record in construction in part on the low skill levels and poor language comprehension of migrant workers. In an attempt to the improve literacy standards, the organisation established French language teaching on large building sites, complementing classes already provided by the state and local associations (FPB 1969, p. 57) (Fig. 3.3).

The building industry was confronted with other work-related problems that had less to do with the specific nationality or immigration status of the workforce, but which were entrenched in deeper,

---

[5]See, for example, interview with Lakhdar in Hervo and Charras (1971, p. 84).

Fig. 3.3 Workers prepare the formwork for the reinforced concrete foundations of a section of the voie express rive droite (now voie Georges-Pompidou) that would pass below pont de Tolbiac at quai de Bercy, Paris, 11 May 1965 (*Source* Coll. Pavillon de l'Arsenal, cliché DUVP)

long-term practices of construction. Workers of all backgrounds complained about the general working conditions of the Parisian building site, which could be disorganised, dirty and dangerous. Workers' unions regularly highlighted the precarious employment practices of the industry: abhorrent conditions, long working hours and insufficient pay and benefits (*PC* 1963e; *JO* 1964; *Le Plâtrier* 1964). The temporary nature of construction work and the sometimes isolated location of building sites caused continual challenges for workers to find housing and make travel arrangements (Fau 1989; Mazé 1993). Low wages and a shortage of affordable housing meant builders struggled to find basic reasonable accommodation. Tens of thousands of workers and their families lived in shanty towns known as *bidonvilles* in the suburbs of the Paris region; self-built shacks located on wasteland or building sites housed factory and construction workers who were largely, although not exclusively, labour migrants from southern Europe and the Maghreb. Living conditions in the overcrowded encampments were dismal: water supply was precarious and could freeze up in winter; wood stoves posed a fire hazard and resulted in numerous child deaths; insufficient drainage and sanitation gave rise to disease (Bozzi 1970; Hervo and Charras 1971; Hervo 2001). Charitable organisations, trade unions and the political left attacked the government for creating an abundant, cheap labour force to build a modern nation; but, living in squalid conditions, this population did not benefit from any of the "progress" of the "civilised" nation ("Collecte organisée pour les habitants des bidonvilles" 1963; *Le Monde* 1966b; *PC* 1968). The sometimes miserable experience of construction workers, on site and at home, provided a dark contrast to the glamorous representations of building sites made available to the general public. Little wonder that industry leaders and state television gave scant attention to the workers they encountered: their plight would considerably diminish the apparent splendour of French construction.

Workers' Unions and the Politics of Construction

Taking the remit of construction work to its limits, the Parisian building site became a stage where politics could be enacted and debated. Working to counter many of the public representations of construction, trade unionism among construction workers transformed the building site into a space of political activism, informed principally by the rhetoric of the *Confédération générale du travail* (CGT, or General Confederation of Labour), which was closely affiliated with the *Parti communiste français* (PCF, or French Communist Party). CGT construction unions sustained an uncompromising critique of industry leaders and media representations that ignored the difficulties faced by builders. The *Fédération nationale des travailleurs du bâtiment* (FNTB, the national construction workers' union) believed these falsely celebratory imaginations of the building site served Gaullist nationalism and disregarded the urgent needs of workers. The CGT and PCF campaigned fiercely for a "strong" and "independent" French nation, but opposed de Gaulle's ruling political movement, which the communist organisations claimed was "fascist" (Eloi 1963, p. 4) and monopolistic (*PC* 1963c). CGT construction unions organised anti-government rallies from building sites (*L'ouvrier des travaux publics* 1962); condemned police brutality towards Algerians in France (Labrousse 1961); organised anti-war movements (AD SD 1966); and campaigned for international justice for workers (*Lettre fédérale* 1963).

The CGT made explicit connections between these wider international political campaigns and the everyday organisation of French building sites. The unions accused employers and the government for increasing labour immigration in order to establish a cheap, abundant "army" of industrial workers in order to drive down wages and create division among workers of all nationalities (*PC* 1963a). The FNTB opposed the establishment of the European Economic Community, which had exposed small French constructors to open competition from large European contractors: the Paris regional branch of the union declared the common market a "war" on the working class (*PC* 1963b). The CGT disagreed with the frequent government regularisation of illegal migrant workers, a practice the Paris construction workers' bulletin branded an "anti-working-class and anti-national policy" (*PC* 1963c). A representative of the FNTB, meanwhile, denounced the naturalisation process, whereby migrant families become "fix[ed] on our land" (Carli 1968, p. 5). This ugly and sometimes racist discourse upheld the narrow assumption a worker in France should be *white working class*, and cast an uncomfortable shadow over the more positive CGT campaigns to improve the conditions of migrant workers. Despite its fundamental opposition to immigration, the CGT agreed it had the duty to defend the rights of workers of all backgrounds, especially migrant construction workers, who the union saw as "victims" of the social and economic inequalities of capitalism (Eloi 1963; *Lettre fédérale* 1963). The CGT established partnerships with representatives of Algerian workers to improve working conditions and literacy (*PC* 1963d), and developed links with unions in other countries in the attempt to establish an international workers' movement. Unions faced resistance from employers who sometimes disciplined non-French workers for participating in strike action—falsely declaring it was illegal for foreign workers to lay down their tools (Vinhas 1968). Large contractors maintained a strong anti-union position, blacklisting militants (Languillier 1997, p. 15) and dismissing foreign trade unionists (*PC* 1964). The battle for justice on the building site would be fought long and hard, especially as non-unionised workers continued to fulfil the appetite of a labour-hungry industry.

The architectural profession, for its part, faced existential challenges from problems related to construction work. Both the changing technological and human aspects of the building site undermined conventional architectural practice, which, until 1968, was still dominated by the classical training of the beaux-arts tradition, mixed with the strict rhetoric of modernist architectural and urban theory. The booming construction industry in France was not necessarily good news for architects, whose education rarely prepared them for the design of complex structures that employed reinforced concrete or steel construction. The largest construction projects relied more on engineers than architects; technical design offices, led by engineers who supervised surveyors and draftsmen to produce calculations and diagrams, were far more commonplace on building sites than architects and their drawings (Hefford 1963). A number of large housing developments such as Les Courtillières in Pantin and Sarcelles-Lochères had been designed by prominent architects—Émile Aillaud and Jacques-Henri Labourdette, respectively—but their involvement in these schemes did little to forge good general public opinion about architects, following criticism in the popular press for the perceived design failings of certain large suburban housing estates, which some commentators believed contributed to social problems (Chomette 1962).[6] One of the most prominent cases that exposed the difficulties faced by the architectural profession dealing with changes in construction practice came when three architects, alongside contractors and site managers, were found guilty of contributing to the deaths of twenty workers when a twelve-storey building

---

[6]For detailed analysis of the design and social experience of suburban housing estates, see Tellier (2007) and Cupers (2014).

collapsed during construction in Paris in 1964 (Paskins 2013). For this project, the architects had been contractually obliged to supervise the assembly of a complex part-prefabricated steel construction system, but one of the project architects explained in the criminal trial that his education at the École des beaux-arts had not prepared him to work with recent construction techniques (Théolleyre 1967). Some architects believed they faced threats from elsewhere in construction. The *Union nationale des syndicats du personnel des cabinets d'architectes, des bureaux d'études d'architecture et d'urbanisme* (a CGT architects' trade union) identified the rise of property developers as a danger to the architectural profession, asserting that developers who sidestepped architects reduced the construction of buildings solely to a form of financial speculation, and "devalued" French architecture (*Lettre fédérale* 1966). Although architects had some legitimate concerns about the challenges they faced from private developers, architects sometimes appeared more concerned with defending the status of their profession than acknowledging the human roles involved in the fabrication of buildings. The architectural profession, like the construction industry, showed little interest in looking after the building workforce it was reliant on.

The state of the national construction industry had long been a measure of economic strength in France: in 1850, Martin Nadaud, a builder and politician, had famously coined the phrase "*Quand le bâtiment va, tout va*", declaring that when construction is working well, all is well in society (Rustenholz 2003, p. 9). But even the post-war French construction boom could not dissimulate the social injustices of an exploitative industry. The relentless growth in construction during the early 1960s eventually stalled and was interspersed with bouts of sluggish employment, regardless of the traditionally quiet winter seasons (*Le Monde* 1966a, c). Job insecurity resulted in a decline in living standards, especially among migrant construction workers, and the spread of impoverished housing and *bidonvilles* became shocking evidence of the unequal distribution of wealth resulting from economic growth (Couvreur 1966). The predicament of workers constructing modern France was at odds with the promise of a progressive future announced by projects like the RER beside the *Arc de Triomphe*.

Architecture, Cities and Construction Work

It is possible that because of the construction industry's apparent lack of attention to the everyday challenges presented by building work, architectural and urban historians have tended to neglect the role labour plays in the production of the city. This is by no means unique to France but is evident across eras and geographies. The reluctance of critics and historians to look beyond pristine façades when interpreting newly completed buildings can be traced to at least Sigfried Giedion (1928/1995)—the historian and spokesman of architectural Modernism—who like his compatriot Le Corbusier would have wanted everyone to believe the most complex structures could now simply be slotted into place; factory-produced parts and machinery would eliminate any individual actions of construction workers. Subsequent histories of architecture and urban planning have focused on the theories and designs of canonical male architects, with little regard for considering what actually happens on building sites. The architectural humanities have more recently, however, begun to acknowledge the role of work in the design and construction professions, resulting in interdisciplinary research that reflects the complexity of cities and the entanglement of the built environment with culture, politics and society. But doing so, as the research that led to this chapter on construction in Paris discovered, raises methodological challenges that confront deep-rooted disciplinary practices.

Architectural and urban history has traditionally been closely implicated in design practice—often written by architects or planners, and drawing on visual sources produced by the professions to inform its interpretations. As the case of urban transformation in Paris shows, the professional's view of a profession can be strikingly at odds with everyday realities and experiences on the building site, attempting to eliminate the presence and input of workers. Examining construction and building workers can offer rewarding new knowledge about the production of architecture and the city. If cities are complex spaces of flows and processes (Wakeman 1997, p. 265), the building site is a reminder that the city is in a constant "state of becoming" (Merriman 2007, p. 3).

The research in this chapter owes much to the influence of Linda Clarke's (1992) classic text *Building Capitalism* that sought to understand how even the lowest paid worker fits into the production of the city—as much, if not more than the traditional professions responsible for "the exchange, distribution, circulation and consumption of the completed buildings" (4). The actions of developers and speculators have dominated discourses about urban development, in both the past and present, but, argues Clarke, the role of builders in creating the city is just as important. Examining changing employment practices, wages, training and skill requirements in eighteenth- and nineteenth-century London, Clarke analysed how labour is deeply implicated in construction and urban design. Architectural and urban history that foregrounds the building site produces an interpretation of the city that emphasises social organisation over architectural style or typological development.

Construction history is a relatively young discipline that situates itself between architecture, engineering and technology, promising to combine the material study of structures with an analysis of the social consequences of construction (Lorenz 2006). Yet despite its interdisciplinary strengths, the field has tended to focus on construction techniques or design typologies: just 8% of articles published by the journal *Construction History* between 1985 and 2014 focused on labour, trades and the professions, and only 1% of articles covered questions about training and education (Addis 2014). Meanwhile, if a focus on the construction worker brings a social dimension to the understanding of architecture and the city, caution should nevertheless be exercised to avoid heroic insider accounts that foreground builders' supposed "masculine" attributes of physical strength, pride and hard work (Applebaum 1981). Proposing histories of urban transformation in the Paris region, some accounts from within the planning and construction professions suffer from similar blind spots, emphasising the political manoeuvring and middle management of these industries, rather than the specific role of workers on the ground (Vibert-Guige 1993; Guyard 2003).

Adrian Forty (2012) has explored the role of human labour in the production of modern architecture through a study of "concrete and labour" (pp. 225–251), but he acknowledges that historians have insufficient knowledge about the "human aspects" of construction due to the limited documentary evidence available about how workers dealt with processes such as the pouring of concrete (p. 230). Architectural and urban history therefore needs to embrace alternative research methods to bring the worker back to the long-abandoned building site. In the case of France in the 1960s, if the trade press provided little information about workers, other non-professional sources provided evidence of the lives and activities of builders. Newspaper articles, trade union records, television and radio programmes, films and even songs form an archive of everyday life and can also identify construction projects not normally deemed worthy of study, challenging the dominance of prestigious, monumental projects in the architectural canon. Oral history and interviews are also key sources to pursue to reconstruct worker activities—a challenge for architecture to turn away, momentarily at least, from visual sources to embrace ethnographic methods of the social sciences (Wall 2013). This is complex, interdisciplinary research that reflects the interconnected systems of cities and society. Architectural and urban historians wishing to gain closer knowledge of construction workers in the twentieth century can learn much from the lead taken in the fields of industrial relations and labour history (Palladino 2005), but the challenge is to use this knowledge to inform a spatial and morphological understanding of the built environment. As Illaria Giannetti (2015) shows with her study of innovative techniques in the building of Italy's *Autostrada del Sole*, major construction projects result in quantities of written documentation and visual sources produced by construction companies, patent offices and government agencies, which provide researchers with a wealth of material to work with, bridging gaps about the relationship between design and construction processes.

What contribution, therefore, can a historical study of the Parisian building site make in the effort to acknowledge and even improve construction work? A good place to start would be to critique the marginalisation of construction workers that is apparent in history as much as it continues in many parts of the world today. Concentrating on expanding the construction industry through modernisation and international ventures, French business and political leaders chose to disregard the workers who were powering this growth. Builders of many different origins and backgrounds, however, faced derisory working and living conditions in an industry seemingly untroubled by the exploitation of its most important assets: the workers. Retrieving evidence of Paris's former building sites makes apparent the otherwise overlooked social, material and political aspects of architecture, urbanism and construction. The transformation of Paris relied on the daily actions of poorly paid workers in precarious employ, who toiled in sometimes dangerous and disorganised sites before returning exhausted to unsatisfactory housing. Yet few official records of construction projects properly acknowledged the individual and collective roles of workers building the modern city.

To present-day readers, a critical examination of Paris's urban transformation fifty years ago is likely to draw uncomfortable parallels between similar cases of exploitation of construction workers in different countries today. Attempts by architects, construction and marketing professionals to dissimulate the role of workers in luxury developments are now perhaps even more widespread than the popular representations of construction in 1960s Paris that eliminated the people building the city. Architectural history and criticism need to disrupt more than ever the triumphalist narratives propagated by industry practices and publicity that edit out the presence of workers. The history of exploited builders is painfully repeating itself across the globe, not least in high-profile projects in the Gulf states, such as the construction of world cup stadia in Qatar, where migrant workers have reportedly been forced to live in "squalid accommodation", received delayed payments and have had their passports illegally confiscated (Gibson 2016). The gap between the photogenic celebration of architectural showcases endorsed by global investors and the deadly realities of the people building them—a Nepalese worker died on average every two days constructing infrastructure in Qatar in 2014 (Gibson and Pattisson 2014)—highlights an uncomfortable coexistence of social extremes that collide within a single project. In these cases, excessive wealth founded on financial speculation or national promotion ultimately speculates on the individual lives of builders.

How can the conditions of 1960s Paris still be replicated across the world in ever more extreme cases of modern-day slavery? Is the global construction site condemned to remain one of the most exploitative and dangerous places to work? The case of the Paris building site suggests that construction can only improve with an acknowledgement of professional problems and a shared responsibility that involves developers and architects, as much as historians and workers. Although some architects deny having any responsibility for the deaths and problems of construction workers realising their designs (Riach 2014), this attitude sets a dangerous precedent for further professional neglect. Beyond the question of who is doing the building, architects face other important ethical considerations including the source of finance for projects or the ecological provenance of construction materials (Deamer 2015). Realising designs into built forms requires knowledge of the complex global networks of capital, supplies and labour that are inherent in the construction industry (Wilson et al. 2015), even more than in Paris of the 1960s. Corruption and exploitation might face more obstacles if architects led the initiative to work with more responsible clients and contractors. Architectural and urban historians also share some of the blame for the apparently endless replications of poor construction practices that should have been long assigned to the past. Architectural history has taken too long to acknowledge its poor record of understanding the position of workers in construction projects, motivated in part with its concern to celebrate architectural theories and practices rather than the activities of building.

Workers, finally, have an important role to play in resisting their ongoing exploitation by learning how to form effective worker movements while facing increasing repression from contractors or gangmasters. In France, trade union organisation and campaigning made significant steps in the 1960s to create long-term improvement in pay, safety and professional reputation in the construction industry, improvements that have been matched in many countries. There still exist, however, considerable problems with the exploitation of undocumented workers in Paris (Nguyen 2017) and instances of blacklisting of trade union activists in the UK (Macalister 2016). If the wider industry continues to grow on the misery of its workers, builders will need to take more active roles than ever to improve their plight (Fig. 3.4).

Fig. 3.4 Construction of the Maine-Montparnasse development of offices, housing and new mainline railway station, 27 September 1967 (*Source* Coll. Pavillon de l'Arsenal, cliché DUVP)

The French construction industry in the 1960s was deeply implicated in the predicament of building work, but it tended to bury its head in the sand, as if it was trying to ignore the material and social realities of work, including employment practices, education, training and skills, the implications of industrialisation and above all labour migration. Similarly, representations of building work produced by the mainstream media paradoxically diminished or eliminated the role of human labour in construction processes. Through a focus on construction activity in the Paris region in the 1960s, this chapter has explored the roles labour played in the development of the city and analysed how these were tied up with geopolitical debates about national identity and migration. The activities of workers both on and off the building site reveal that architecture, planning and urban design are never the result of a single, visionary designer but are always produced by the complex interplay of different actions, trades and people that sometimes reach beyond the domains of design and engineering professionals. Actors in urban development in the 1960s, as much as today, include bureaucrats, engineers, financiers, manufacturers, media organisations, non-design professionals, politicians, residents, trade union militants and workers. Each of these groups has the responsibility to construct a better world to work in.

References

Addis, B. (2014). The contribution made by the journal *Construction History* towards establishing the history of construction as an academic discipline. In J. W. Campbell et al. (Eds.), *Proceedings of the first conference of the construction history society* (pp. iii–x). Cambridge: Construction History Society.

AD SD (Archives départementales de la Seine-Saint-Denis, Bobigny). (1966). Fonds photographique du journal *L'Humanité*. Manifestations pour le paix au Vietnam, 81Fi/727, 3 February, 30 June.

Allain, Y. (1966). Paris de demain (newsreel). *Chroniques de France*, Pathé Cinéma. Forum des Images, Paris.

Applebaum, H. A. (1981). *Royal blue: The culture of construction workers*. New York: Holt, Rinehart and Winston.

Bozzi, R. (1970). *Les immigrés en France* (film). Parti communiste français. Forum des Images.

Bullock, N. (2002). *Building the post-war world: Modern architecture and reconstruction in Britain*. London: Routledge.

Carli, F. de. (1968). Rapport de la commission exécutive fédérale. *Lettre fédérale, 162* (July–August), 3–9.

Chateau, R. (1967). *Nouvelles techniques d'injection et de paroi moulée appliquées à la construction* (film). Solétanche/RATP. Forum des Images.

Chomette, H. (1962). Fiche au dossier des grands ensembles. *Techniques et Architecture, 23*(1), 114.

Clarke, L. (1992). *Building capitalism: Historical change and the labour process in the production of the built environment*. London: Routledge.

Collecte organisée pour les habitants des bidonvilles (radio news report). (1963). *Inter-Actualités de 7h15*, RTF, 28 January. Fonds radio. Insitut national de l'audiovisuel, Paris (INA).

Couvreur, J. (1966). Bidonvilles et sous-prolétariat urbain. Series of six articles, *Le Monde*, 1, 2, 4, 5, 7, 8 June.

Cupers, K. (2014). *The social project: Housing postwar France*. Minneapolis: University of Minnesota Press.

Deamer, P. (Ed.). (2015). *The architect as worker: Immaterial labor, the creative class, and the politics of design*. London: Bloomsbury.

DGDP (Délégation générale au district de la région de Paris). (1965). *Schéma directeur d'aménagement et d'urbanisme de la région de Paris*. Paris: La Documentation française illustrée.

Diefendorf, J. (1993). *In the wake of war: The reconstruction of German cities after World War II*. Oxford: Oxford University Press.

Eloi, J. (1963). Rapport général d'activité et d'orientation. *Lettre fédérale*, 94 (March), 3–12.

"Enquête annuelle d'entreprise: Bâtiment et travaux publics, chantiers 1968–1969." (1973). Supplement, *Statistiques de la construction*, 8 (September), 27.

Fau, A. (1989). *Maçons au pied du mur: Chronique des 30 années d'action syndicale*. Montreuil: FNTC-CGT.

Forty, A. (2012). *Concrete and culture: A material history*. London: Reaktion.

FPB (Fédération parisienne du bâtiment et des activités annexes). (1966). *L'Emploi des étrangers dans le bâtiment*. Paris: Éditions H. Vial.

FPB. (1969). *L'Entreprise du bâtiment et la formation de son personnel*. Paris: Éditions H. Vial.

Franc, R. (1971). *Le scandale de Paris*. Paris: Grasset.

Gall, P. le, (1965, October 18). Les constructeurs se plaignent: on nous oblige à truquer! (includes an interview with Raymond Camus). *Le Parisien libéré*.

Giannetti, I. (2015). The Italian story of Ferdinando Innocenti's tubolar scaffolding (1934–64). In B. Bowen, D. Friedman, T. Leslie, & J. Ochsendorf (Eds.), *Proceedings of the fifth international congress on construction history* (Vol. 2, pp. 177–184). Chicago: Construction History Society of America.

Gibson, O. (2016, March 31). Migrant workers suffer 'appalling treatment' in Qatar World Cup Stadiums, says Amnesty. *Guardian.*

Gibson, O., & Pattisson, P. (2014, December 24). Death toll among Qatar's 2022 World Cup workers revealed. *Guardian.*

Giedion, S. (1928/1995). *Building in France, building in iron, building in ferroconcrete.* Santa Monica, CA: Getty Center for the History of Art and the Humanities. Originally published as *Bauen in Frankreich, Bauen in Eisen, Bauen in Eisenbeton.* Leipzig: Klinkhardt & Biermann.

G. M. (1965, October 5). L'activité du bâtiment augmentera de 5% par an en moyenne jusqu'en 1970. *Le Monde.*

Guyard, J. (2003). *Evry, ville nouvelle, 1960–2003: la troisième banlieue.* Evry: Espaces Sud.

Hefford, J. J. V. (1963). *The French building industry.* London: Royal Institution of Chartered Surveyors.

Hein, C., Diefendorf, J. M., & Yorifusa, I. (Eds.). (2003). *Rebuilding urban Japan after 1945.* Basingstoke: Palgrave Macmillan.

Hervo, M. (2001). *Chroniques du bidonville: Nanterre en guerre d'Algérie.* Paris: Seuil.

Hervo, M., & Charras, M.-A. (1971). *Bidonvilles: L'enlisement.* Paris: François Maspero.

"Ils sont trois millions de travailleurs étrangers en France." (1964). (Includes an interview with Pierre Bideberry, director of the ONI). An episode in *Sept jours du monde,* ORTF, 12 June.

JO (Journal officiel de la République française: Débats parlementaires). (1963). Préfabrication dans la construction. 22 June, 3660.

JO. 1964. "Protection des ouvriers du bâtiment." 28 November, 5706.

Labrousse, L. (1961). La guerre d'Algérie: L'urgence d'en finir! *Le Plâtrier, 92* (October), 1.

"La main-d'oeuvre dans le bâtiment et les travaux publics." (1974). Supplement, *Statistiques de la Construction,* 11 (July–August).

Languillier, J. (1997). *Une chienne de vie.* Bourges: Jules Languillier (self-published memoir).

"Le général de Gaulle visite le Paris de l'an 2000" (newsreel). (1966). *Les Actualités françaises,* 18 May. Fonds de presse filmée. INA.

Le Monde. (1965a, November 10). Le marché du travail: Le bâtiment emploie le plus grand nombre d'immigrés algériens.

Le Monde. (1965b, May 20). Les délegués ont évoqué la situation sociale et notamment celle des jeunes.

Le Monde. (1966a, January 5). Le president de la Fédération du bâtiment souligne le 'climat d'incertitude' qui règne dans la profession.

Le Monde. (1966b, June 10). Les indésirables…ou l'implantation de foyers de travailleurs étrangers.

Le Monde. (1966c, April 23). Trois mille chômeurs algériens et africains dans le bâtiment parisien.

Le Monde. (1967, January 24). Une enquête de la Caisse des ouvriers du bâtiment confirme la précarité du niveau de vie de certains pensionnés.

Le Plâtrier. (1964). Dans les entreprises. 114 (August–September), 2.

Leridon, D. (1969). *Austerlitz 70* (film). SNCF. Forum des Images.

"Les nouvelles halles de Rungis" (television news report). (1966). *JT 20H*, ORTF, 24 October. Fonds national de télévision. INA.

Lettre fédérale. (1963). Solidarité aux travailleurs du bâtiment d'Allemagne fédérale. 93 (February), 11.

Lettre fédérale. (1966). Résolution. 137 (April), 2.

"L'industrie du ciment." (1966). *Lettre fédérale*, 139 (June), 4.

Liotard, M. (2007). *Le Havre 1930–2006: La Renaissance ou l'Irruption du Moderne.* Paris: Picard.

Lloyd Thomas, K., Amhoff, T., & Beech, N. (Eds.). (2016). *Industries of architecture.* London: Routledge.

Lorenz, W. (2006). From stories to history, from history to histories: What can construction history do? *Construction History, 21,* 31–42.

L'ouvrier des travaux publics. (1962). 10 (January–February), 1.

Macalister, T. (2016, May 9). Blacklisted workers win £10m payout from construction firms. *Guardian.*

McDonald, J. R. (1969). Labor immigration in France, 1946–1965. *Annals of the Association of American Geographers, 59*(1), 116–134.

"Malraux à Pontoise" (television news report). (1967). *JT 20H*, ORTF, 12 December. Fonds national de télévision. INA.

Mazé, P. (1993). *Les bâtisseurs: Chronique de 150 ans de luttes sociales.* Paris: FNTC-CGT/Scandéditions.

Merriman, P. (2007). *Driving spaces: A cultural-historical geography of England's M1 motorway.* Oxford: Blackwell.

Nguyen, L., (2017, January 27). Travail Dissimulé: Les Forçats du Métro se Battent Pour Etre Payés. *L'Humanité.*

Nye, D. (1994). *American technological sublime*. Cambridge, MA: MIT Press.

Ockman, J. (2015). Foreword. In P. Deamer (Ed.), *The Architect as worker: Immaterial labor, the creative class, and the politics of design* (pp. xxi–xxvi). London: Bloomsbury.

Palladino, G. (2005). *Skilled hands, strong spirits: A century of building trades history*. Ithaca: Cornell University Press.

Paskins, J. (2010). The representation by French television of building construction work in and around Paris during the 1960s. In N. Bose & L. Grieveson (Eds.), "Using Moving Image Archives," a special e-book edition, *Scope: An Online Journal of Film and Television Studies*, *17*, 82–100.

Paskins, J. (2013). The boulevard Lefebvre disaster: A Crisis in construction. *Architectural Histories, 1*(1), 1–15. https://doi.org/10.5334/ah.ax.

PC (Paris-Construction). (1963a). Appel aux travailleurs immigrés. 1 (April), 6–7.

PC. (1963b). Le chancelier ou le maçon? 1 (April), 2.

PC. (1963c). Qui fait venir la main d'oeuvre étrangère? 3 (October–November), 6.

PC. (1963d). Travailleurs algériens participez en masse à la campagne d'alphabétisation. 3 (October–November), 7.

PC. (1963e). Vestiaire ou taudis? 2 (July), 5.

PC. (1964). Les travailleurs immgirés dans la lutte. 6 (November), 4–5.

PC. (1968). Pour les travailleurs déplacés: Mettre fin aux conditions d'hebergement inhumaines. 22 (December), 2.

Pilat, S. (2014). *Reconstructing Italy: The Ina-Casa neighbourhoods of the postwar era*. Farnham: Ashgate.

PL (Le Parisien libéré). (1966, January 1–2). Discours télévisé du général de Gaulle.

"Pose dernier élément pont Charenton" (television news report). (1968). *Journal de Paris*, ORTF, 14 October. Fonds national de télévision. INA.

"Procédé Camus: Béton préfabriqué en usine." (1962). *Techniques et Architecture, 22*(5), 151.

Ragon, M. (1968). *La cité de l'an 2000*. Tournai: Casterman.

"Rétrospective Paris-chantiers" (television news report). (1966). *Journal de Paris*, ORTF, 30 December. Fonds national de télévision. INA.

Riach, J. (2014, February 26). Zaha Hadid defends Qatar World Cup role following migrant worker deaths. *Guardian*.

RICS (Royal Institute of Chartered Surveyors). (1975). *The French building industry*. London: Royal Institute of Chartered Surveyors.

Ritter, K. (Ed.). (2012). *Soviet modernism 1955–1991: Unknown history.* Zürich: Park Books.

Rustenholz, A. (2003). *Paris ouvrier: Des sublimes aux camarades.* Paris: Parigramme.

Tellier, T. (2007). *Le temps des HLM, 1945–1975.* Paris: Autrement.

Théolleyre, J.-M. (1967, November 22). Les trois architectes en cause soutiennent qu'ils n'avait pas à contrôler la bonne execution d'un procédé technique. *Le Monde.*

"Trois mille tonnes en mouvement." (1966). An episode in *Les coulisses de l'exploit,* ORTF, 18 May. Fonds national de télévision. INA.

Vernholès, A. (1965, October 5). L'industrialisation de la construction permettrait une baisse sensible des prix des logements. *Le Monde.*

Vibert-Guigue, A. (1993). *Au temps des chemins de grue: Chronique des années de béton.* Paris: Éditions des Alpes.

Vinhas. (1968). Extraits des interventions des delégués. *Lettre fédérale,* 162 (July–August), 28.

Voldman, D. (1997). *La Reconstruction des Villes Françaises de 1940 à 1954: Histoire d'une Politique.* Paris: l'Harmattan.

Wakeman, R. (1997). *Modernizing the provincial city: Toulouse, 1945–1975.* Cambridge, MA: Harvard University Press.

Wall, C. (2013). *An architecture of parts: Architects, building workers and industrialisation in Britain, 1940–1970.* London: Routledge.

Weiss, S. (2015). Frozen assets: Photography, time, and labor on the construction site. In B. Bowen, D. Friedman, T. Leslie, & J. Ochsendorf (Eds.), *Proceedings of the fifth international congress on construction history* (Vol. 3, pp. 577–584). Chicago: Construction History Society of America.

Wilson, M. O., Carver, J., & Baxi, K. (2015). Working globally: The human networks of transnational architectural projects. In P. Deamer (Ed.), *The architect as worker: Immaterial labor, the creative class, and the politics of design* (pp. 144–158). London: Bloomsbury.

4

Liberating the Semantics: Embodied Work(Man)ship in Construction

Rikard Sandberg, Christine Räisänen,
Martin Löwstedt and Ani Raiden

Introduction

What theoretical insights may we derive from individual stories of how subjects embody their social and cultural worlds? How do they craft their selfhood to accommodate or resist subject positions inscribed and ascribed to them by established discursive practices? What kinds of agentic embodiment are enacted to resist such hegemonic discourses in order for change to be effectuated? These questions are currently debated in the social sciences within several fields, such as work studies

R. Sandberg
Chalmers University of Technology, Gothenburg, Sweden

C. Räisänen (✉) · M. Löwstedt
Department of Architecture and Civil Engineering,
Chalmers University of Technology, Gothenburg, Sweden
e-mail: christine.raisanen@chalmers.se

A. Raiden
Nottingham Business School, Nottingham Trent University,
Nottingham, UK

© The Author(s) 2018
D. J. Sage and C. Vitry (eds.), *Societies Under Construction*,
https://doi.org/10.1007/978-3-319-73996-0_4

(Virno 2004; Sennet 2009; Flemming 2014; Frayne 2015); organisation studies (Hassard et al. 2000; Holliday and Hassard 2001; Wolkowotz 2006; Dale and Latham 2015); and gender studies (Butler 1990, 1993, 2004; Kondo 1990; Nelson 1999; Brewis and Sinclair 2000; Denissen 2010; Smith 2013; Raiden 2016) to name only a few. Often, these debates tend to be theory driven and therefore high on the ladder of abstraction.

Research on work in organisations has been criticised for neglecting to account for human bodies and corporeal differentiation in enacting organisational realities (Acker 1990; Leder 1990; Collinson and Hearn 1994; Hassard et al. 2000). For instance, Acker (1990) has argued that the stereotypical employee, as portrayed in both organisational theory and practice, seems to be a disembodied and universal organisational actor who lacks bodily attributes, such as gender, sexuality and affect, and who performs abstract jobs. In reality, Acker argues, organisations are pervaded by representations of masculine bodily attributes that discursively and materially frame how we intrinsically come to understand and make sense of structures, processes and positions in organisations. Human bodies have come to constitute paradoxical phenomena; they are central to inquiries into organisational life yet are obscured in the very same inquiries. Bodies are taken for granted, remaining 'an absent presence' (Shilling 2012). In spite of the growing body of research on work in organisations, how individuals, especially lower-level managers and workers, deploy their bodies to navigate subject positions made available to them warrants further research. Situated studies of individuals' embodied enactments of work practices could enlighten (interfere with or contradict) theory (Holliday and Hassard 2001, p. 7). In this chapter, we provide situated empirical examples of one individual's perceptions of her socially embedded body and the active choices she makes to appropriate and occupy viable subject positions made available in the given dominant masculine discourse. Against the backdrop of largely Anglo-Saxon studies of women doing construction, this chapter contributes experiences from a Scandinavian context.

Using a narrative approach and narrative analysis, we have collected twenty-nine work-life stories of site managers in several large- and middle-sized construction organisations to investigate individual affective experiences and corporeal responses to situations arising in daily work and life practices. Over the last two decades, the construction literature has shown that the job pressures and responsibilities of site managers have been diversifying and increasing, impacting their health and possibly the quality of their work (e.g. Djerbani 1996; Haynes and Love 2004; Lingard and Francis 2004; Styhre and Josephson 2006; Dossick and Neff 2011; Mäki and Kerosuo 2015). Furthermore, tensions between the organisation and site managers often arise since for the latter experience-based intuition and judgement acquired in projects and used in daily work to solve problems have much higher currency than do the standardised management systems advocated by the organisation (e.g. Dubois and Gadde 2002b; Styhre 2011, 2012; Sandberg et al. 2016). We wanted to know how construction managers embody their work and non-work roles and responsibilities vis-à-vis themselves, their co-workers and the organisation. It is important to note here that we were not targeting gender issues; these were inevitably and naturally addressed by the respondents themselves, either explicitly or implicitly.

In this chapter, we focus on one respondent's body and its performance[1] in an attempt to bring out, and separate, the perceived experiences and emotions of a physical body-in-context from prevalent discursive representations. Our presentation revolves around the life story of Mona, a Swedish senior site manager, whose life story reveals how elusive, paradoxical and complex identity work is for managers (female and male) in a male-dominated industry. In accordance with Holliday and Hassard (2001), we view a body as multiplicitous, 'a site through which competing and contestant discourses of power and resistance are played out' (see also, e.g., Hassard et al. 2000;

---

[1]Butler distinguishes between 'performance' and 'performativity'. The former she defines as a bounded act by a deliberate subject while the latter is iterative enactment of norms which precede, constrain and exceed the performer's 'will' or 'choice' (Butler 1993, p. 24). We argue that the performer's enactment cannot only be reduced to discursive prescription, but that he/she in engaging with the discourse may resist or intervene by appropriating available subject positions (Smith 1988; Nelson 1999).

Butler 1990, 2004; Mol 2003). In particular, we examine the relationship between strongly entrenched masculine corporeal inscriptions of an ideal (normal) body and embodied performances of construction work as these unfold in practice. Our analysis highlights nuances and complexities concerning the interplay between embodied subjective agency in deploying gendered strategy choices in situated contexts. In doing this, we follow Smith (1988) and Nelson (1999) and show how a subject may be constituted through dominant discourses *and* simultaneously may be reflexive *and* intervene in the process. It is the *how* and the *why* of these interventions that we wanted to understand. We thus contribute empirically and theoretically to further understanding of gendered embodied processes in the social sciences and in construction management. Empirical data from construction practices are scarce in social science theorising in general, and in the embodiment literature in particular. We show that construction is a rich and fertile empirical site for challenging and expanding social science theorising on the body and on work.

Contested Bodies in Construction

Construction is historically and prevalently a male-dominated industry. It is a project-based, site-specific practice where conditions of uncertainty, complexity and immovability of the constructed product justify a decentralisation of organisational authority to the projects (Dubois and Gadde 2002a). Construction projects are heterogeneous configurations, gathering a wide number of stakeholders from different spheres, professions and organisations that come into the project at different stages and in predetermined sequences. The activities that unfold on construction sites have often been described as chaotic and complex (Cicmil and Marshall 2005; Ness 2012), constituting an 'ad hoc' environment in which unanticipated situations constantly emerge and militate against formal planning and standardisation of work activities.

In the construction literature as well as in practice, bodies are essential. The physically and often harsh conditions in combination with a strong craftsman tradition have favoured qualities such as autonomy, self-sufficiency, endurance and roughness (Hayes 2002; Applebaum 1999; Thiel 2007; Denissen 2010). On this 'Planet Construction' (Creed 2000, p. 181), a masculine 'construction tribe' reigns to endure and deal with the mud and dirt, and to 'get things done' in spite of limited resources. These capabilities are strong signifiers of competence, and they underpin the ubiquitous sociocultural representation of a masculine rational mind and strong body (e.g. Applebaum 1999; Thiel 2007; Styhre 2011; Ness 2012; Pink et al. 2012; Ajslev et al. 2016). 'Real work' (Ness 2012, p. 661) on Planet Construction is site based, entailing lengthy commutes, prolonged working hours, challenging work conditions and pressure to complete projects on time. The culture and physicality of the work therefore are often viewed as focal points around which other conditions are negotiated and reinforced (Paap 2006; Ajslev et al. 2016; Wolkowitz 2006). Culture and identities have been argued to be the outcomes of a range of interacting material, symbolic, affective and practical conditions, leading construction scholars to describe these conditions as entangled (Ajslev et al. 2016; Wolkowitz 2006). This implies that the constitution and (re)production of construction-worker identities (and behaviours) cannot be reduced to any one condition, but must be understood against the background of entangled and emergently (re)productive material and ideological conditioning in time and local spaces, and that identities are fluid and changing (Dale and Latham 2015; Brewis and Sinclair 2000; Barad 2003).

The hubs of communication, coordination and orchestration of activities and interfaces in the industry are the site managers, who must be attuned to the many different cultures, processes and tools of the different interacting professions and trades (e.g. Dossick and Neff 2011; Styhre 2012; Mäki and Kerosuo 2015). Styhre (2012) depicted the work practices of site managers as a 'muddling through', illustrating a reactive pattern of skilfully solving problems as these arise and trying to be everywhere at the same time. 'Muddling through' according to Styhre is *required* by site managers in order to cope with the manifold tasks within their area of responsibility. This kind of 'crisis management' together with its proclivity for 'overwork', combined with a 'paternalistic leadership', is a strong contributory force to the perpetuation and reproduction of masculine embodiment of an ideal (normal) construction worker: one who is autonomous, authoritative, self-reliant, single minded, tough and hard (Styhre 2011, p. 947; see also Brewis et al. 1997[2]).

Barring the 'paternalistic' (which we come back to later), there is something disturbing in this argument, but it may be so that this something is a question so obvious that it has remained unarticulated and therefore obscured: What is it that makes the six embodied qualifiers enumerated above signifiers of masculinity? Over time, these collocated and reinforcing meanings have been appropriated, (re)conceptualised and reproduced in masculine discourses of management and work to constitute essentially male qualifiers. They are so deeply ingrained in the culture and structure of construction that their genesis is a black box—a matter of fact—and the discursive point from which theorising takes off. No matter how sophisticated the theorising, the underlying fixed binary of male–female drives behaviour and thinking such that 'autonomous', 'authoritative', 'self-reliant', 'single-minded', 'tough' and 'overworking' are ubiquitously culturally inscribed as male.

In the construction literature, the hegemonic image (material as well as symbolic) of masculinity in the industry combined with the masculine 'nature' of the work is rarely problematised (for recent critical inquiries, see, e.g., Chan 2013; Rumens 2013; Smith 2013). If we want to work towards 'undoing gender' (Butler 2004) and strive 'for gender to matter less' (e.g. Raiden 2016, p. 520), we need to untangle these semantics and (at least try) to liberate them from their gendered burden. Similarly, the word 'caring', for example, has been appropriated by feminist discourses and likewise needs to be liberated.

The construction literature so far has provided ample evidence that the socially articulated masculine norm is indeed the ideality against which bodies are measured, and, more importantly, against which these

---

[2]Brewis et al. (1997, p. 1280) depicted the prevailing modern(ist) ideal of 'real men' using a similar enumeration of masculine-appropriated qualifiers, namely 'logical, argumentative, potent, masterful, hardworking, and incrementally progressive'.

bodies measure themselves (e.g. Paap 2006; Denissen 2010; Smith 2013; Raiden 2016). Weighted against the conventional, socially articulated representation of women's bodies (and minds) as messy, uncontrollable and 'leaky', it is not so difficult to understand the inculcated notion that women deviate from the established 'norm' (Butler 1990, 1993, 2004; Holliday and Hassard 2001). Following Brewis et al. (1997), we would also argue that men who do not uphold the ideal construction masculine body are similarly prone to derogatory scrutiny.

Yet, in defiance of the prevailing hegemony, the number of women in construction-oriented trades and professions is increasing (e.g. Whittock 2002; Wall 2004; Watts 2007, 2009; Ness 2012; Braundy 2011). Moreover, there is empirical evidence that they enjoy the work, and in spite of their 'deviating bodies' (Lindgren and Packendorff 2006, p. 857), they are doing a good job (e.g. Smith 2013; Raiden 2016). Creed (2000, p. 183) reported that:

> [A] resilient and influential minority of women does exist within the industry, who *are* likely to hold alternative viewpoints. This group is likely to increase as the present young cohorts of women professionals grow older and become more 'cynical', thus contributing to build-up of critical mass. (italics and quote marks in original)

As a consequence, bodies in construction, especially the taken-for-granted masculine bodies, may be becoming contested terrain (Holliday and Hassard 2001). It is timely to turn to unfolding practices and revisit working bodies to further explore how dominant discourses, local practices *and* subjective agency intersect to organise and regulate bodies through corporeal strategies and subjective sense making. Skeggs (1997, p. 82) pointed out that the body is the 'physical site where the relations of class, gender, race, sexuality and age come together and are embodied and practiced'. This 'coming together' is in constant flux, constantly transformed, as one's body is 'be[ing] cast, always, outside oneself, Other to oneself' (Butler 2004, p. 148). It is this 'being cast' (being formed) outside oneself, and the labour required to reflexively negotiate viable subject positions to accommodate alterity *as well as* one's own representation of self that we think needs to be problematised. To quote Butler (2004, p. 20) again:

In a sense, to be a body is to be given over to others even as the body is, emphatically, 'one's own,' that over which we must claim rights to autonomy. (quote marks in original)

Bodily autonomy, as Butler (2004, p. 21) argues, is double edged: it is agent and instrument at the same time. Before we can lay claim to our own body, it has already been claimed and inscribed by the social and cultural worlds to which we belong; in other words, our bodies are primordially inscribed by various dominant discourses in our environment. So according to Butler, a woman entering construction is inherently 'given over' to the dominant discourse of the 'real man' construction ideal, i.e. to a male appropriation of semantics. Following Butler's argument, claiming autonomy remains a fantasy; the I/gender of the woman entering construction is thus intrinsically 'undone' by the ruling discourse. If she embodies any of the masculine-appropriated qualifications, she is impinging on the territory of the 'Other' and will be humbled; and if she upholds appropriated feminine qualifications, she is likewise humbled. Similarly, men who do not conform to the reigning ideal are equally chastised. Thus, gender is performed continuously through a stylised repetition of acts where the repetition covers its own genesis (Butler 1990), and the appearance of substance becomes confused with the meanings that reproduce it. In Butler's terms then, masculinity is not something that construction can be said to *have*, but it is continuously done and re-done through discourse. Through the postmodern tendency of treating subjects, bodies and actions as always constituted in and through discourse, the subject is robbed of its agency, and the socio-material contingencies of the contexts in which the performances unfold are ignored (e.g. Lloyd 1999; Nelson 1999; Barad 2003). We therefore collude in covering the genesis!

Smith (1988) and Nelson (1999) critiqued such an ontology, arguing that a subject is not only constituted through and in dominant discourses, *but* may simultaneously engage reflexively and reflectively with the discourse and thus create a subject position. This position may be transformative for the subject and for the dominant discourse.

As Spicer et al. (2009, p. 544) suggest: 'through being made performative, discourses create spaces where we are able to rework them'. This reworking of the already 'cast body' requires awareness and reflexivity on the part of an agentic subject in an effort to liberate the masculine or feminine semantic appropriation of qualifiers. This could be a way of claiming autonomy.

Research Design

The empirical material for this chapter revolves around the story of Mona, a senior site manager who works in a large construction company in Sweden, where she has been employed for the major part of her thirty-year-long career. Mona is well respected and considered a highly skilled and successful site manager in the organisation, an image that has been confirmed by both her superiors and subordinates. Mona is currently managing a mid-sized construction project which is running over budget and is causing her anxiety.

Mona's story is part of a larger study on the work and life of site managers in the construction industry, where we hitherto have conducted interviews with twenty-nine site managers. Of these twenty-nine managers, Mona's interview has been followed up as a sub-case of its own, where we zoom in and follow her life story over the course of several years. So far, we have interviewed Mona on two occasions with a one-year interval. We see Mona's story as representing an 'insightful example' (Alvesson and Deetz 2000) of body work rather than conveying an unconditional truth about the nature of construction work. Mona's story problematises taken-for-granted assumptions about construction work, masculinity as well as bodies and their relationships (Alvesson and Sandberg 2013). Ellis and Bochner (2000, p. 744) argued for the importance of the personal story in research, suggesting that it 'offers lessons for further conversation rather than undebatable facts'. The question to be asked is therefore not how well the personal story represents a certain 'reality', but what consequences the story produces (p. 746). In the interviews with Mona, we made extensive field notes on her verbal as well as corporeal expressions: 'spoken words are, strangely, bodily offerings: tentative or forceful, seductive or withholding, or both at once!' (Butler 2004, p. 172). We wanted the interviewee to speak freely so that her 'being would become elaborated' in new ways through our conversation (Butler 2004, p. 173). We tried to create a situation where speech would not be controlled, but rather that the conversation would be a 'mode of doing something together and becoming otherwise' (ibid.).

> The speaker may be relaying a set of events in the past, but the speaker is doing something more: the speaker, in speaking, is presenting the body that did the deed, and is doing another deed at the same time, presenting the body in action. (Butler 2004, p. 172)

During the interview, Mona was first asked to describe her career path to date. After these preliminaries, she was encouraged to talk freely about her work and work role. 'Free' storytelling has been suggested as an appropriate interview technique for the purpose we had in mind, where the interviewee's personal story is allowed to evolve, and in which underlying assumptions and beliefs guide the conversation (Clandinin and Connelly 2000). Our prompts were therefore open-ended; we wanted her to tell us about her workdays, how she generally went about planning and managing site activities, what issues arose and how she dealt with them. A narrative analysis was used on the data, entailing that the various accounts and fragments of accounts were identified and sorted in storylines that linked to the overall experience of her work-life practices (Czarniawska 2004). We found five related story lines, or bodies, that together created a narrative that helped us make sense of Mona's embodied reality.

Here, we try to render Mona's voice, words and embodied performances as closely as possible even when we paraphrase and summarise her utterances by using quote marks for her words. We are aware that nuances inevitably disappear in translation; however, since one of us is bilingual (a native speaker of English and Swedish), we feel that this risk is mitigated.

Introducing Mona: Multiple Embodied Enactments

Driven by her proclivity and interest in 'hard' subjects, e.g. engineering and construction, Mona consciously chose a traditionally male-dominated education and career path from an early age. She loves her job, the many challenges she has to face every day, the hard work and most of all the freedom she feels she is afforded. In the eyes of her superiors and her team, she is a competent and respected manager and leader: a 'determined lady' as one of them said. Running through her storylines are strains of hegemonic masculine discourse, within which, as she shows, she crafts her work identity. In doing this, she engages with the dominant discourse reflexively, purposefully seeking available spaces from which she can intervene and disrupt the prevailing discourse (e.g. Nelson 1999).

Like all narratives, Mona's is imbued with ambiguities and contradictions which underpin as well as undermine cornerstones of her credo: a pursuit of autonomy and freedom; a need to be in full control; a desire to be 'a competent site manager'; and a fantasy of a 'better' future for herself. Her beliefs and job situation bear similarities with the life stories of other women who have chosen male-dominated career paths, depicted, for example, in Braundy (2011), Raidan (2016), Smith (2013) and Brewis and Sinclair (2000). Like Mona, these women seem to embody a similar attitude of 'ok! that's the way it is, so let's get on with it!' And they do, deploying an array of gendered masculine and feminine strategies, both reinforcing the masculine prevailing norm and attempting to feminise it. Mona's story contributes to these studies by providing a Scandinavian perspective and adding nuances to performativity theorising.

In this section, Mona's, bodily enactment of four masculine norms as showcased in her life story is presented: a reactive body; an omnipresent body; an absent body; and an autonomous body. These bodies are explored as different modes of embodied performances that emerged and interlaced throughout her narrative of her work and non-work life.

We argue that these multiple bodies are enacted productions and iterations of the masculine norm and discourse of her work environment (following Styhre (2011) and Brewis et al. (1997)). From this perspective, Mona's performative bodies emulate, repeat and perpetuate normalised realities and reinforce their normalisation power (Butler 2004, p. 53). In this respect, she embodies the abstract pre-constituted subject Butler has been criticised of representing. However, we also show how Mona mobilises her deviating body to disrupt and interfere with normalised discursive gendered preconceptions to reflexively, reflectively and purposefully change them by using culturally available meanings and reworking these through purposive enactment in her day-to-day activities.

In the following, we start by introducing Mona as we perceived her when we met her. We then let her speak for herself and showcase her various embodied actions and reactions as she navigates and negotiates spaces and times. In this section, we examine her deployment of masculine-appropriated qualifications or masculine-gendered strategies, and in the next section, we show how she reflectively mobilises her female deviating body.

Drawing on our field notes, we start by introducing Mona.

We are sitting in the reception area waiting and contemplating numerous pictures of impressive infrastructure and buildings in-the-making. An interesting exhibition of symbolic capital, which seems to be common to construction firms. We wait … and we wait … Mona finally sprints toward us, markedly stressed, explains quickly that she is going through a turbulent period and can only give us 40 minutes (of the promised 90)! Curtailing the interview, she says, is better than cancelling it. We proceed through a long corridor toward a meeting room. During the walk, we chat. Here and there in the corridor, small groups of men converse with each other, a woman exits a room carrying documents. She nods at us. The men are uniformly dressed in chequered blue or black, lumberjack-style shirts over dark slacks; some are wearing a short-sleeved pullover. Mona too is dressed in sombre colours: a black silk blouse over dark women's trousers and short-heeled pumps. Her thick, unruly shoulder-length hair is let out. Mona is of average-height and sturdily built (she readily jokes about her 'overweight'). She walks with determination toward one of the group of men. Taking no notice of us, she embarks on a discussion with them. They speak loudly, using technical jargon. We observe. What we see are four bodies in animated and attentive conversation, gesticulating, nodding, frowning, listening, and gazing at whoever is momentarily speaking, which is mostly Mona. With her input, they all seem to reach a mutually satisfying decision, after which they abruptly shift to rather derogative joking about the misbehaviour of a sub-contractor. Mona seems to be in full command of the situation: she exudes an aura of self-confidence and assertiveness. Her speech has a completely different clang in this group: her local dialect and idiom have become more pronounced, and she has lowered her pitch. Although this conversation only lasts a couple of minutes, it provides a valuable snapshot of workplace (and gender) interaction as it unfolds.

The Coping and Reactive Body: Problem-Solver and 'Firefighter'

Mona describes her work role as free and very rewarding, yet at the same time says it is 'extremely' demanding. She evokes excessively long workdays, periodically at a relentless pace and with an excessive workload. She explains that she is in a particularly hectic phase of the project right now, practically working non-stop. The project is running over budget, and she anticipates more problems. She discloses that in spite of all her experience, she is finding it difficult to remain unfazed when 'everything is tumbling down'. During such phases (which seem to be frequent), she tends to go into 'firefighting mode', she says, where she constantly must prioritise. She admits that at such times she deals with everyday management activities reactively, in an ad hoc manner.

As a result of such chaotic situations, Mona perceives her work situation as fragmented; she feels split between all the different activities and responsibilities of the site. She emphasises that there simply is not enough time to do all the things that she perceives are expected of her. This perception of insufficiency, which she shares with other women managers depicted in the literature (Paap 2006; Denissen 2010; Smith 2013; Raiden 2016), may be in the eye of the beholder rather than articulated expectations of the organisation. As far as we could make out in Mona's case, colleagues and superiors bore her much respect and appeared to view her as a very skilled and competent manager who required neither monitoring nor control.

However, recent studies of site managers (e.g. Djerbani 1996; Haynes and Love 2004; Lingard and Francis 2004; Styhre and Josephson 2006; Dossick and Neff 2011; Styhre 2012; Mäki and Kerosuo 2015) report on the increasing work burden of this level of managers, men and women, who are having to take on additional administrative, economic and people-related responsibilities. These are neither articulated in their work descriptions nor do the managers receive any training for these add-on work tasks.

There is no such thing as a typical workday, Mona declares. The only thing that can be considered as given and recurrent is that she starts work early and that she dashes around resolving an endless stream of emerging issues. She finds it hard (near impossible) to foresee in advance all the problems that will need to be dealt with throughout a regular workday. She therefore needs to always be prepared to be interrupted; she also needs to be able to improvise.

The kind of work ideology that Mona articulates is not unique for construction. In their invective of capitalism's corporate work ideology, Cederström and Fleming (2012) bleakly depict work metaphorically as a 'virus' (p. 12) that is colonising our bodies. We are no longer able to switch work off:

> I have been getting up at 4.30 am and leaving work at 7 pm, and going to bed at 9 pm. How does one count time with such a schedule! During the weekend, I sit with the budget. I easily work a 100-hour work week …

Working a 100-hour week at this pace would indeed, as Cerderström and Fleming argue, result in a 'Dead Man [sic] Working'. Mona rationalises that the only reason she endures such a regime is her knowledge that 'occasional' calmer periods will follow, and then she can recover:

> The fact that it goes up and down like this is both the charm and the downside of this job. I mean if it were this hectic all the time, I would never be able to put up with it … because I am <u>really</u> tired right now. I just need to keep on struggling … until it gets calmer.

However, when she describes 'calmer' periods, they do not strike us as very calm at all. Even in these periods, her work routine is interrupted by a constant stream of unanticipated problems at different locations on and off site, requiring her presence. This may be anything from deliveries not showing up on time, lack of available resources and personnel, or sudden unexpected geo-technical problems on the site. Mona again stresses that being a site manager is a very demanding role that requires an 'extraordinarily' committed and stress-hardy individual, who has to be present as much as possible:

> You practically have to be superman to manage all the demands …

In line with the site managers in Styhre's (2012) study, Mona typically enacts the pragmatic 'muddling through' of site managers, whose jobs revolve around formal planning of operational work activities on site and skilful solving of unforeseeable problems that constantly interrupt the plans by demanding on-the-hoof and immediate solutions. Despite the toll such muddling through has on her body, she takes obvious pride in her problem-solving skills, her perseverance and her desire to be present as much as possible on site. Yet, she feels she is 'insufficient', a theme which runs as a leitmotif through her storylines, and which she acknowledges generates stress-related symptoms, such as anxiety, and physical problems such as hip pain and blood clots.

The Omnipresent Body: Everywhere at the Same Time

As already mentioned, Mona arrived late for the interview. She seemed flurried, and after the improvised discussion with colleagues in the corridor, she needed a few minutes to catch her breath. A natural follow-up for us then was to ask her if it is common for her to rush like this between different locations. Her answer comes as a loud laugh, whereupon she explains that not only is it common, she feels as though she 'does it for a living'.

Mona's body is constantly in motion, even as she is sitting here in a small meeting room facing us across a smallish round table. Her body twitches and moves around in the chair. She is very present yet seems ready to spring up as though she is in a starting block. A central theme in Mona's story is the importance of mental and bodily presence; she bemoans her physical inability of being present in all the places and on all the occasions she feels it is expected of her. The most demanding expectations, though, seem to be those she puts on herself. She tells us: 'the minute I wake up in the morning, my body tells me I need to be somewhere else'. Hence, she always aims to be the first one to arrive on site in the morning.

> I always try to be the first one in every morning. I know this is something that 'the guys' appreciate, and I know that they talk about it. (quote marks expressed with her hands)

This is one of the very few times in our conversation that she alludes to her flesh and bone male colleagues rather than to an ideal site manager. It is an interesting example of impression management, supporting the notion that to be accepted, women have to perform better than men. She also explains that she wants to arrive early because it is the only time of the day she can work undisturbed and focus on budgeting and planning, i.e. the principle formal tasks of a site manager.

She confesses that during some of the most hectic periods, she has spent the night in her office cubicle in order to save the time spent commuting from home by car.

> It has happened that I have slept at my desk waiting for a morning meeting … with work charts and drawings as covers to protect me from the cold.

She rationalises that sleeping at her desk is her way of alleviating some of the stress which accumulates during periods of excessive overwork. She also is quick to point out that in 'ordinary' circumstances she values spending time with her family. Observe that she does not say that 'she does spend time!'

Mona has not taken a single day off ill in several years. It does not matter how ill she is; she makes it a point to show up at work to carry out the most critical tasks. She recalls one occasion in the final phase of a project when she contracted a stomach flu:

> I can say this much, I had my second blood clot last year and contracted salmonella at the same time. I had a bucket with me on my round of the site and had to run off into the woods for privacy ... that's the way it works! So, I go to work, and I have a bucket. It functions. I'm not all that ill!

For Mona, presence no matter what is a means of enacting physically and mentally what she sees as the representation of a 'competent site manager'. Episodes like the bucket and sleeping at her desk are obviously satisfying to her because they both embody and symbolise corporeal proof that she is managing *her* project no matter what the circumstances. In contrast to many previous accounts of women managers and workers in construction trades (e.g. Braundy 2011; Smith 2013; Raiden 2016), it is interesting to note that Mona does not explicitly articulate gratification in her ability to do a *man's job* or satisfaction in handling traditional masculine tools. Furthermore, we do not discern in her utterances the vulnerability and uncertainties that are said to be masked under women's deployment of masculine discourse (e.g. Raiden 2016). Mona exudes self-confidence in her managerial abilities and measures her performance against her perception of a 'competent site manager' rather than her male colleagues. A self-assessment questionnaire carried out among Swedish female and male managers in construction showed that the women considered themselves to possess requisite managerial competences, scoring high in decision-making and action (Arditi et al. 2013), which may explain the difference. Mona also demonstrates an existential awareness of her work and life choices which sensitises us to the recursive relationship between dominant discourse and reflexive socially embedded subjects at the micro-level (Nelson 1999).

Mona's need to always be present, in control and 'on top of things' reflects a masculine idealisation of heroic leadership and discourse, which accords with accounts of women in construction by, e.g., Smith (2013) and Raiden (2016), and reflects the authoritative, self-reliant, single-minded, tough and hard requisite qualities of site managers in construction (Styhre 2011; Brewis et al. 1997). Part of a woman's body work in this environment, therefore, can be said to involve managing representations of self in relation to others. Comparing the men's utterances as quoted in Styhre and Josephsson (2006) and in Styhre (2012) as well as those of our twenty-eight male respondents, we perceive that the men's language is less emphatic, less affective, when they describe the trials and tribulations of their job. Their body work, although mentioned, seems less foregrounded. This perception warrants comparative investigations of men and women's talk about their work. For now, we suggest that the men in construction do not need to emphasise that which is 'normal', i.e. that which has already been semantically appropriated as masculine; for them the meanings of the aforementioned qualities have already been normalised and are unmarked. As Holliday and Hassard (2001, p. 5) state: 'the body becomes invisibilised through the techniques of normalising bodies', which then entails that men do not need to foreground their body work since to them it is normal and natural. This would explain the critique directed towards the invisibility of bodies in the organisation literature. Based on this argument, we could argue that Mona's (and other women's) foregrounding of corporeal enactments reflects her need to disrupt and claim (i.e. liberate) the masculine semantic appropriations of meanings as applying to site managers in general, both male and female.

The Absent Body: The Downside of Presence

From the accounts so far, Mona seems to work all the time. We therefore probed the other facets of her life. Does she have a life outside of work, and what does that life look like?

As alluded to already, Mona's high expectations on herself engender a constant emotional sense of insufficiency.[3] This becomes the leitmotif of her story; it appears in many forms to mar the various bodies she enacts

---

[3]Note that we make a clear distinction between uncertainty and insufficiency: the former we see as denoting insecurity of some kind while the latter refers to a lack of some kind, most often due to external factors.

on the job. It manifests in the constant feeling that she articulates of needing to be somewhere else, solving some other problem. Her presence is bodily; her mind is one step ahead. As she herself points out, such a work pattern is intrinsically unsustainable and unhealthy. Since human bodies cannot be physically present in all places and at all times, the expectations and demands she imposes on herself inevitably entail absences. Paradoxically, it is the absences that cause her the most anxiety and stress. Absence preys on and in her presence, causing much of her sense of insufficiency. From the accounts so far, Mona seems to work all the time. We therefore probed the other facets of her life. Does she have a life outside of work, and what does that life look like?

Mona confides that the only reason she has been able to work at such a pace is because she has neglected her family, her friends and her own well-being. She talks of these aspects of life as the 'collateral damage' of her work life.

> I don't have any alone time. I definitely don't have time to meet friends.
> My family I hardly see at all ... so these bits are the collateral damage.
> I never go to the cinema; I don't have time for such things.

She has one daughter, having made the conscious decision to abstain from having more children. Her husband, she says, works at the same pace as she does, and the spouses share the household shores and caring, which is a common scenario among full-time double-earner families in Scandinavia today (e.g. Raiden and Räisänen 2013). Despite her reducing family to collateral damage, she nevertheless says that her family is a very important element in her life. She points out that she tries to be present in her daughter's life, and she feels that she and her daughter have a strong bond. However, the insufficiency leitmotif in this context is overlaid by an affectual undertone of compunction. It surfaces when she describes how she and her daughter sit together watching children's programmes on television in the evenings, but she is not watching; she is working on her laptop. She is present in body, but not in mind. As she is depicting this scene, her body posture, gestures and facial expressions shift. She becomes somewhat subdued, stops twitching and bows slightly forward with her arms crossing her chest. It is the only time in our interview that we get a glimpse of a possible vulnerability. In the next moment, though, she straightens her body and declares that she is happy with the choices she has made. Raiden (2016) and others have pointed to the difficulties full-time working women in construction have in balancing work and family life, and how many, if not most, are childless and/or single. In this chapter, we do not examine the work and family problematics, but here too further comparative research is warranted.

Mona also admits that her work pace has cost her bodily well-being. Besides often recurring periods of exhaustion, she laughingly refers to her overweight as another price she has to pay for wanting to keep 'on top of work'.

> I never have the time for physical training. That is why I don't lose weight. Now though I just have to get started because I have developed a bad hip due to my overweight.

This preoccupation with her weight is an interesting detail in her story. Brewis and Sinclair (2000) see women's relations to their bodies predominantly manifested as a feeling of lacking, often connected to body shape and appearance, mirroring the image of one's 'outside-in' reflection as compared with sociocultural expectations. This feeling of lack is also highlighted in the studies of women in construction cited above. In Mona's case, this feeling of lack is no doubt one reason she alludes to her weight, which she also mentions earlier on in a joking aside. Coming back to her utterance above, she rationalises her overweight as yet another collateral damage albeit in different terms, absence from the gym, as well as pointing to consequences of her work regime on her body's well-being.

Yet, in spite of all the detrimental aspects she articulates, she nevertheless feels that she is coping with the stresses that her job engenders, in both her work life and her home life. Indeed, what seems to bind all the fragments in her story into a consistent narrative—providing us with interpretative cues—is a persistent rationalisation of her situation: She is where she is by choice; she has learnt to cope with stress; her job gives her satisfaction; she can exploit her gender; she weathers periods of overwork and exhaustion to 'enjoy' the relative calm that inevitably follows. She also has a strong bond with her daughter and is there for her, at least in body.

An interesting facet in her story, which we have noticed in our interviews with men as well, is a penchant towards invoking a future respite. For example, Mona mentions that 'she lives for her holidays when she is able to put work out of her mind'. Yet, earlier on, she has admitted that she remains available, even in her free time. This rationalisation of an imaginary 'future perfect' (Clegg et al. 2006) can be seen as more than mere rationalisation for our benefit. We believe that she constructs a fantasy of calm periods and free time to motivate her self-induced overwork (in this respect, see Berlant's (2011) notion of 'cruel optimism'). This phenomenon remains under-researched within the field of work studies and may explain why there is scant resistance to the current escalation of the neoliberal work ideology of the capitalist era (Fleming 2014; Frayne 2015). Indeed, the work ideology has a strong grip on construction workers, which organisations maximise on. So, for example, the organisation encourages site managers like Mona who 'are in control of their site' and 'do not require monitoring' and silently affirm the prevailing site-manager ideal.

We asked Mona whether she ever envisages changing jobs to something less demanding. She acknowledges that she often thinks of quitting and acquiring a more sustainable life situation. Part of the reason why she does not do so is not knowing what she would do instead. Another reason is that she perceives herself as irreplaceable due to her competences and experiences of managing certain types of construction projects (see Arditi et al. 2013). She is also in doubt whether she 'would be given the same amount of freedom in any other organisation'.

It seems that the freedom Mona perceives she has managed to acquire and maintain in the organisation provides the motivation needed for her to continue working in the current position. That this freedom may be elusive, and that it may cost too much, she is reluctant to admit to us, or, as we suspect, to herself.

The Autonomous Body: Elusive Freedom

Mona says she has the freedom to influence her work in the directions she perceives as most meaningful and satisfying for her and her project team. She is also able to resist or refuse tasks she feels are unworthy, uninteresting or unrewarding.

> As long as I can work with what I want in the way I want, I enjoy working here. Today I am certainly in such a position. There are tasks that I am not interested in, and there is <u>no way</u> I am going to perform them.

As we dig deeper into this story line, the energy in the room seems to intensify. Mona speech becomes more assertive as does her whole demeanour. For a brief moment, we get a feel of Mona the site manager, the 'determined lady'. We ask about the boundaries of her freedom, to which she elaborates that she has three core criteria which she must satisfy and in which she must perform well. She must achieve 'good' results (cost, time and quality), be able to maintain her staff, and satisfy the customer. If she performs well in all three areas, she is free to shape her role and organise her time. She describes these three performance criteria as forming a triangle constituting the outer boundaries of her freedom as a site manager. She perceives these boundaries as being very generous and finds great satisfaction in the freedom she perceives she is given. This freedom is important to her; it seems to symbolise her autonomy, and it generates an enormous pride and satisfaction at project completion, when she can say: 'I built that!' This affective ownership of a concrete 'monument' resonated in all our interviews with site managers, and has been noted in all the studies of construction workers hitherto cited. It is probably what makes this industry unique. Contrary to the bleakness of endlessly working to work (Cederström and Fleming 2012), in construction there (still) is an actual (unique) physical monument that lives on after its creators are gone. We witness this human trait in monuments built centuries ago, where the name of carpenters or bricklayers is inscribed in the wood work or the stones.

However, autonomy and freedom are inherently intertwined with an overall unsustainable work situation, which is also prevalent in the role. Mona, for example, takes on tasks and responsibilities that are not part of her formal work description, resulting in further strain on her already heavy workload. Here, interestingly, she contradicts her previous assertions of being readily able to refuse tasks. She admits that she finds it difficult to delegate certain tasks and responsibilities to her subordinates even though she knows that she should. She explains that her body needs to be in control and 'on top of things'. It is not that she does not trust her subordinates, she says, but: 'the tasks are important and I believe that I can perform them better'. The problem, as we see it, is that too many tasks seem to 'feel important', rendering Mona's work situation unsustainable. Flemming (2014) suggested that: 'the more embodied expertise and independence that workers have, the less free they are'. This certainly applies to Mona's case. The more tasks she takes on, the more she proves to herself she is in control and independent, and the more others rely on her to 'firefight' for them, etc.... Could this (evil) spiral be an example of unfolding 'cruel optimism?' As a consequence, the freedom she so strives to embody is also elusive. It does not result in her actually being free, but rather it carries the seed of continuous and growing feelings of frustration, stress and insufficiency that risk undermining her role, and more importantly, her self-worth in the long run.

So far, we have seen how Mona embodies what Styhre (2011) asserts are 'necessary' masculine-gendered strategies to muddle through effectively on site. We also obtain a better understanding of the work ideology in construction. In the next section, we discuss Mona's embodied enactments from the perspective of her deviating body, showing how she attempts to claim back and re-appropriate established masculine meanings.

Deviating Body: Where Women Dare to Tread

Mona of her own accord introduces and speaks candidly about the gendered nature of construction work. She explains that being a woman with a body that deviates from that which is not only predominant, but also normalised within the industry can sometimes be advantageous.

She can strategically use her gender, her marked body; it provides her leverage that her male colleagues, the unmarked (normal and natural) bodies, are denied.

> I often feel that being a woman is an advantage here … <u>yes, really</u>! Because there are so few women in this industry, men are so scared to step on our toes. [...] I have learned to use this in order to get my ideas through and shape my work role the way I want it. (our underlining to mark strong emphasis)

Mona perceives that being a woman in a male-dominated environment gives her advantages which enable her to better manage and enact demands and workload. In the quote above, she seems to anticipate disbelief from us and quickly counters with an emphatic 'yes, really'. (Or maybe her assertion was in fact triggered by one or both of our bodily display of disbelief.) Moreover, the 'yes really' foregrounds her awareness of the power and prevalence of the constitutive masculine norm, and at the same time her appropriation of an available (possible) subject position from which she can craft her own identity (Nelson 1999).

She recounts how, over the years and through experience, she has become sensitised to different kinds of male reactions to her otherness, and she has learnt to use these reactions to her advantage. For example, she explains that for craftsmen and managers in the industry, 'women on site are incongruous elements; they neither possess the competences nor the authority requisite for the job', therefore, 'stating that I know what concrete is suffices to impress male colleagues'. In this somewhat sarcastic utterance, Mona shows reflexive awareness of the constitutive power of a dominant discourse, and how she avails herself of a subject position that opens up as she engages with the discourse.

However, the advantage she may gain comes at a price, which she also reflects on critically. Men, she says, tend to respond with a condescending 'pat on the head' in recognition of her demonstration of 'competence and authority', and accept her, but only conditionally. She thus acknowledges that she indeed is a part of the in-group, yet and at the same time, due to her female body, remains conspicuously deviant.

Mona provides further insight into the complex mix of gendered behaviours identified among women in male-dominated workplaces in her metaphor of the 'pat on the head'. Women use masculine-gendered strategies to show competence and authority, but from the point of view of men co-workers, at least from Mona's reflections, she remains a 'surrogate' male, (an)'other', and as such elicits a mix of disbelieving admiration with a kind of condescension. The latter may be a self-preserving reaction on the part of men to a perceived threat (Kondo 1990; Braundy 2011). We think that more empirical work needs to be done to make visible these underlying culturally produced psycho-social reactions.

Over the several decades that Mona has worked in the industry and in the company, she has seen many women come—and then leave—because they have refused to accept the symbolic, patronising 'pat'.

> If you cannot ignore it [the pat] and see the advantage in capitalising on and exploiting the men, then I think you would find it difficult ... if you are a feminist who strongly sticks to equality principles, expecting to be treated in a certain way, then I think you would find it bloody hard!

Mona, on the other hand, symbolically and bodily 'turns the other check', using masculine-gendered strategies to disrupt normalised meanings. She early on in her career perceived the possibility of a space opening up in the dominant masculine discourse, and she learns how to exploit it in an attempt to penetrate and disrupt gendered embodied responses. She describes how attitudes of male colleagues and project stakeholders shift from initial aloofness to a kind of perplexity as she continues engaging with them. Most often when she meets with new clients or sub-contractors, a moment of 'blankness' arises. This, according to her, is the moment it takes her interlocutors to overcome their apparent confusion of being faced with a manager who is also a woman:

> There is a sudden tangible blankness in the air ... I feel it when I enter the room and am introduced as their [my team's] site manager ... and in how the people react. <u>It is so very obvious! You</u> would probably not feel it, but for me, who has experienced it a <u>long</u> time, I <u>know</u> that is what it is ... it is more [their] astonishment that I <u>know things</u> and am able to <u>do things</u> that I am not expected to know or to do.

Mona becomes very animated as she speaks of these experiences; obviously the recollection still causes frustration, and as we perceived it, hurt pride. The tenor of her voice rises, and facial expressions and gestures indicate that these recollections are releasing pent-up emotions (see underlined phrases and words). The literature so far lacks processual empirical illustrations of the interplay between causes and effects of gendered strategies and body work of the kind Mona manifests here.

Our interpretation of the excerpt above is that it is not Mona's essential deviating body which provokes this reaction in male interlocutors, but what her body represents in terms of mind and knowing. An explanation for the gut reactions that she perceives she elicits is that they are remnants of a culturally subconscious deep-seated Western (gut)belief of men belonging to the public sphere of rational decision-making (first-order function pertaining to the mind) and of women belonging to the private sphere of home and nurturing (second-order function pertaining to the body) (e.g. Martin 1989; Brewis and Sinclair 2000). It is this prejudicial underestimation of her intellectual capacity and the violation of her integrity that upset her the most. Interestingly, even though these recurring episodes are 'so very obvious' to her, she professes doubts whether we would recognise them for what they are. We did not take her up on this statement since in the moment we (first and second author) felt we understood what she meant. In retrospect, we can only speculate. Our interpretation of this instance was then, and still remains, that Mona sees the construction industry as lagging behind in terms of gender equality and equity compared with other industries in Sweden, and especially compared with a university context, which is ours. She may think that in Sweden today the kind of gender discrimination she meets at work is obsolete in other places of work. Mona's story reveals the complex and multifaceted nature of gender at work in construction, and shows how different kinds of male as well as female bodies may be enacted as contextual situations arise.

Being male does not per se engender privilege or advantage. In the interplay of genders in the context at hand, being male may be a disadvantage, if, and seemingly only if, a woman is able to use appropriate and timely gendered strategies. Mona aptly expresses this complexity:

If I'm in the process, gosh ... if one is negotiating or discussing, and embellishes the figures a bit ... one does it all the time ... after all we do live in a world of entrepreneurship ... then one is not questioned in the same way [as men are] ... one is not hindered in the same way.

As already mentioned, Mona has learnt to use the possibilities such instances offer as leverage in her negotiations with clients and colleagues. Her description of how she achieves this is that she: 'simply fills the shoes of the competent site manager', which, she paraphrases as 'not acting primarily like a woman'. Embodying and enacting the 'competent site manager' entails taking initiatives and overtly manifesting know-how and authority when possibilities open up. When she applies this strategy, which often entails not only enacting, but also appropriating masculine embodiment, she feels that she is able to enforce her decisions in situations where her fellow male managers may have more difficulties, for example in negotiations with clients and in refusing tasks she does not want to take on.

We see two reasons for her success in negotiations: first, since men are the unmarked gender, and the qualifiers of a competent manager have been appropriated by the masculine dominant discourse, these qualifiers are 'normal' for men. This we suggest means that to succeed on an individual plane, in the reality of their work negotiations with other men, they have to work harder to make a mark (see, e.g., Hassard et al. 2000; Monaghan 2002). The second reason we suggest is that not only do men fear to 'step on women's toes' as Mona intimates earlier on; it may also be, at least in Sweden, that men are ashamed of that moment of 'tangible blankness' in which their gendered prejudices are unveiled. Mona has learnt to exploit men's fears (Braundy 2011) and fleeting shame by reflexively intervening and engaging with the dominant masculine text (Smith 1988; Nelson 1999). It is maybe this ability that empowers her perceptions of freedom and autonomy.

To be accepted as viable subjects by their 'others', female workers and managers in construction have been shown to largely perform the inscribed gender behaviours by acting as 'surrogate males' (Le Feuvre 1999). In a recent study of a woman site manager's gendered strategies, Raiden (2016) has examined how a female manager draws on a range of feminine and masculine strategies. Drawing on Butler's notion of performativity, Raiden suggests that site-management work for a woman entails negotiating two contrasting discourses: one 'modern, professional and high-tech'; the other 'uniquely tough, masculine and practical' (p. 519). The latter discourse serves to enact the surrogate male and affords women acceptance in the in-group of male colleagues and subordinates; it also abets in reproducing the masculine discourse of the industry. The former discourse of professionalism evokes a managerial identity and allows for the expression of the feminine affective and caring manifestations.

It is interesting to note that both Raiden (2016) and Styhre (2011) evoke caring among site managers, albeit the latter as a *paternal* expression and the former as a *maternal* expression. While caring commonly is perceived as feminine, it is also central to the protective characteristic of the paternal role, which Collinson and Hearn (1994) claim, managers deploy to differentiate themselves from women. From the life stories of the twenty-eight male site managers, most of whom evoked caring in remarkably similar terms to many of the women who talk of mothering and/or caring, we perceive a thin line between paternal caring and the affect-laden mothering illustrated in Raiden. For example, when one of the site managers in our cohort exclaims:

> I think I still feel a craving even though I am getting older. I feel that it is so much fun! Now when young people are coming in and you see how they evolve and you get a kind of paternal role [laughing] … if you are not a dad at home, you can be one at work. Yes, that's how it is. (from our data)

We have difficulties interpreting his genuine affective expression as evoking the paternalistic discourse described by Collinson and Hearn (1994), Styhre (2011) and others. We also note that Mona does not mention mothering or caring in the context of her work. In the same way that we need to liberate terms from their masculine-appropriated meanings, so too do we need to liberate terms like 'caring' from their feminine-appropriated meanings. We strongly encourage further social science-oriented research on discourses at the micro-level in construction-related contexts, for example using a critical-discourse analytic lens.

Closing Remarks

In the work environment of construction, women remain a minority among the operative and managerial workforce on site. Their presence is a deviation from a long-lived, traditional norm, and therefore in itself engenders disruptions (Smith 2013; Raiden 2016) of, and in, normalised behaviours; e.g., men may have to pay heed to how they act, talk and joke; women may feel they have to live up to the masculine ideology by becoming 'surrogate males'. According to Smith (2013, p. 863), the most obvious disruption is women being on a construction site in the first place, 'literally being in the wrong place', doing operative work and managing a team consisting most often solely of men. The assumptions underlying this claim, in our view, undermine the integrity of the career choices of the individual women at hand. Do these women see themselves as 'being in the wrong place'? We doubt that this is so; in fact, for Mona, 'this place' is the only place in which she can envisage herself as being despite the many negative aspects she may evoke. Most of the hardships she describes do not relate as much to her gender as to a strong personal as well as cultural (working class) work ethos. This neoliberal work ideology is fast becoming dominant, and even though it stems from a masculine work ideology, it afflicts all of us (Cederberg and Fleming 2012; Fleming 2014; Frayne 2015; see also Raiden 2016; Sandberg et al. 2016). Problematising the dominant discourse of the neoliberal ideology of work and its effects on embodied work may be a fruitful way of disrupting the gender binary.

To conclude, construction work(man)ship and leadership are historically, culturally and materially/bodily inscribed as masculine (e.g. Paap 2006), and this à priori positioning of masculine attributes (Hassard et al. 2000) is reproduced by both genders in spite of feminising strategies and gender equality regimes and policies. This historical and cultural inscription is so powerfully inculcated in individuals entering these trades already from their education and apprenticeships that personal proclivities, interests and competence require hard work to measure up to the ideal of a site manager.

Mona does not question the values and norms that she embodies. In this respect, we can argue that she colludes in the perpetuation of the norm by 'not behaving like a woman' although she leaves the interpretation of what she means by 'woman' open. This complicity has also been noted in the accounts by Smith (2013) and Raiden (2016). However, Mona also states that living up to the ideal site manager is corporeally impossible, regardless of gender:

> You practically have to be superman in order to manage all the demands…

Superman is indeed the ultimate embodiment of the male ideal worker in this context; 'he' flies to the rescue, solves problems as they crop up, and is to all intents and purposes omnipresent and omniscient, as well as 'seemingly' free. No matter how much she tries, or how much she may desire, she feels she will always be wanting in the eyes of the industry and organisation. More importantly, her body is wanting as seen through her own eyes.

Women in construction may have *potential* situational advantages; however, to avail themselves of these, they have to perform typical traits of the unmarked 'other' *as well as* those of their marked gender. They have to engage with the dominant discourse, i.e. interfere with it by seeking out possible subject positions from which they can claim autonomy by liberating site-managerial qualifiers from their burden of masculine meanings. Experience and reputation does not seem to precede a skilled woman site manager, rather for each new encounter, and often in re-encounters, she has to (re)seek and (re)grab available spaces in the discourse when these arise.

References

Acker, J. (1990). Hierarchies, jobs, bodies: A theory of gendered organizations. *Gender & Society, 4*(2), 139–158.

Ajslev, J. Z., Møller, J. L., Persson, R., & Andersen, L. L. (2016). Trading health for money: Agential struggles in the (re)configuration of subjectivity, the body and pain among construction workers. *Work, Employment & Society.* https://doi.org/10.1177/0950017016668141.

Alvesson, M., & Deetz, S. (2000). *Doing critical management research.* London: Sage.

Alvesson, M., & Sandberg, J. (2013). *Constructing research questions: Doing interesting research.* London: Sage.

Applebaum, H. A. (1999). *Construction workers, USA* (No. 54). Westport, CT: Greenwood Press.

Arditi, D., Gluch, P., & Holmdahl, M. (2013). Managerial competencies of female and male managers in the Swedish construction industry. *Construction Management and Economics, 31*(9), 979–990.

Barad, K. (2003). Posthumanist performativity: Toward an understanding of how matter comes to matter. *Journal of Women in Culture and Society, 28*(3), 801–831.

Berlant, L. (2011). *Cruel optimism.* Durham: Duke University Press.

Braundy, M. (2011). *Men and women, and tools: Bridging the Divide.* Nova Scotia: Fernwood.

Brewis, J., & Sinclair, J. (2000). Exploring embodiment: Women, biology and work. In J. Hassard, R. Holliday, & H. Willmott (Eds.), *Body and organization.* London: Sage.

Brewis, J., Hampton, M. P., & Linstead, S. (1997). Unpacking Priscilla: Subjectivity and identity in the organization of gendered appearance. *Human Relations, 50*(10), 1275–1304.

Butler, J. (1990). *Gender trouble.* London: Routledge.

Butler, J. (1993). *Bodies that matter: On the discursive limits of sex.* London: Routledge.

Butler, J. (2004). *Undoing gender.* Abington: Sage.

Cederström, C., & Fleming, P. (2012). *Dead man working.* Winchester: Zero Books.

Chan, P. W. (2013). Queer eye on a 'straight' life: Deconstructing masculinities in construction. *Construction Management and Economics, 31*(8), 816–831.

Cicmil, S., & Marshall, D. (2005). Insights into collaboration at the project level: Complexity, social interaction and procurement mechanisms. *Building Research & Information, 33*(6), 523–535.

Clandinin, D. J., & Connelly, F. M. (2000). *Narrative inquiry: Experience and story in qualitative research.* San Franscico: Sage.

Clegg, S., Pitsis, T. S., Marosszeky, M., & Rura-Polley, T. (2006). Making the future perfect: Constructing the Olympic dream. In D. Hodgson &

S. Cicmil (Eds.), *Making projects critical* (pp. 266–293). Basingstoke: Palgrave Macmillan.

Collinson, D., & Hearn, J. (1994). Naming men as men: Implications for work, organization and management. *Gender, Work & Organization, 1*(1), 2–22.

Creed, C. (2000). Women in the construction professions: Achieving critical mass. *Gender, Work & Organization, 7*(3), 181–196.

Czarniawska, B. (2004). *Narratives in social science research.* London: Sage.

Dale, K., & Latham, Y. (2015). Ethics and entangled embodiment: Bodies–materialities–organization. *Organization, 22*(2), 166–182.

Denissen, A. (2010). The right tools for the job: Constructing gender meanings and identities in the male-dominated building trades. *Human Relations, 63*(7), 1051–1069.

Djebarni, R. (1996). Impact of stress on site management effectiveness. *Construction Management and Economics, 14,* 281–293.

Dossick, C. S., & Neff, G. (2011). Messy talk and clean technology: Communication, problem-solving and collaboration using Building Information Modelling. *The Engineering Project Organization Journal, 1*(2), 83–93.

Dubois, A., & Gadde, L.-A. (2002a). Systematic combining: An abductive approach to case research. *Journal of Business Research, 55,* 553–560.

Dubois, A., & Gadde, L. E. (2002b). The construction industry as a loosely coupled system: Implications for productivity and innovation. *Construction Management & Economics, 20*(7), 621–631.

Ellis, C. S., & Bochner, A. (2000). Autoethnography, personal narrative, reflexivity: Researcher as subject. In N. Denzin & Y. Lincoln (Eds.), *Handbook of qualitative research* (pp. 733–768). Thousand Oaks: Sage.

Fleming, P. (2014). *Resisting work: The corporatization of life and its discontents.* Philadelphia: Temple University Press.

Frayne, D. (2015). *The refusal of work: The theory & practice of resistance to work.* London: Zed Books.

Hassard, J., Holliday, R., & Willmott, H. (Eds.). (2000). *Body and organization.* London: Sage.

Hayes, N. (2002). Did manual workers want industrial welfare? Canteens, latrines and masculinity on British building sites. *Journal of Social History, 35*(3), 637–658.

Haynes, N., & Love, P. (2004). Psychological adjustment and coping among construction project managers. *Construction Management and Economics, 22*(9), 129–140.

Holliday, R., & Hassard, J. (2001). *Contested bodies*. London: Routledge.

Kondo, D. (1990). *Crafting selves: Power, gender, and discourses of identity in a Japanese workplace*. Chicago: University of Chicago Press.

Le Feuvre, N. (1999). Gender, occupational feminization, and reflexivity: A cross-national perspective. In R. Crompton (Ed.), *Restructuring gender relations and employment* (pp. 150–178). Oxford: University Press.

Leder, D. (1990). *The absent body*. Chicago: University of Chicago Press.

Lindgren, M., & Packendorff, J. (2006). What new in new forms of organizing: On the construction of gender in project-based work. *Journal of Management Studies, 43*(4), 841–866.

Lingard, H., & Francis, V. (2004). The work-life experiences of office and site-based employees in the Australian construction industry. *Construction Management and Economics, 22*(9), 991–1002.

Lloyd, M. (1999). Performativity, parody, politics. *Theory, Culture & Society, 16*(2), 195–213.

Mäki, T., & Kerosuo, H. (2015). Site managers' daily work and the uses of building information modelling in construction site management. *Construction Management and Economics, 33*(3), 163–175.

Martin, E. (1989). *The woman in the body: A cultural analysis of reproduction*. Milton Keynes: Open University Press.

Monaghan, L. F. (2002). Hard men, shop boys and others: Embodying competence in a masculinist occupation. *The Sociological Review, 50*(3), 334–355.

Mol, A.-M. (2003). *The Body multiple: Ontology in medical practice*. Durhum: Duke University Press.

Nelson, L. (1999). Bodies (and spaces) do matter: The limits of performativity. *Gender, Place & Culture, 6*(4), 331–353.

Ness, K. (2012). Constructing masculinity in the building trades: 'Most jobs in the construction industry can be done by women'. *Gender, Work & Organization, 19*(6), 654–676.

Paap, K. (2006). *Working construction: Why white working-class men put themselves—and the labor movement—in harm's way*. Ithaca: Cornell University Press.

Pink, S., Tutt, D., & Dainty, A. (Eds.). (2012). *Ethnographic research in the construction industry*. Abingdon: Routledge.

Raiden, A. (2016). Horseplay, care and hands on hard work: Gendered strategies of a project manager on a construction site. *Construction Management and Economics, 34*(7–8), 508–521.

Raiden, A., & Räisänen, C. (2013). Striving to achieve it all: Men and work-family-life balance in Sweden and the UK. *Construction Management and Economics, 31*(8), 899–913.

Rumens, N. (2013). Queering men and masculinities in construction: Towards a research agenda. *Construction Management and Economics, 31*(8), 802–815.

Sandberg, R., Raiden, A., & Räisänen, C. (2016). Workaholics on site! Sustainability of site managers' work situations? In M. Prins, H. Wamelink, B. Giddings, & K. Feenstra (Eds.), *Proceedings of the CIB World Building Congress 2016.* Tampere: Tampere University of Technology.

Sennet, R. (2009). *The craftsman.* London: Penguin.

Shilling, C. (2012). *The body and social theory.* London: Sage.

Skeggs, B. (1997). *Formation of class and gender: Becoming respectable.* London: Sage.

Smith, P. (1988). *Discerning the subject, theory and history of literature* (Vol. 55). Minneapolis: University of Minnesota Press.

Smith, L. (2013). Trading in gender for women in trades: Embodying hegemonic masculinity, femininity and being a gender hotrod. *Construction Management and Economics, 31*(8), 861–873.

Spicer, A., Alvesson, M., & Kärreman, D. (2009). Critical performativity: The unfinished business of critical management studies. *Human Relations, 62*(4), 537–560.

Styhre, A. (2011). The overworked site manager: Gendered ideologies in the construction industry. *Construction Management and Economics, 29*(9), 943–955.

Styhre, A. (2012). Leadership as muddling through: Site managers in the construction industry. *The work of managers: Towards a practice theory of management* (pp. 131–145). Oxford and New York: Oxford University Press.

Styhre, A., & Josephson, P.-E. (2006). The bureaucratization of the project manager function: The case of the construction industry. *International Journal of Project Management, 24*(3), 271–276.

Thiel, D. (2007). Class in construction: London building workers, dirty work and physical cultures. *The British Journal of Sociology, 58*(2), 227–251.

Virno, P. (2004). *A grammar of the multitude: For an analysis of contemporary forms of life.* Semiotext(e): Los Angeles.

Wall, C. (2004). 'Any woman can': 20 years of campaigning for access to training and employment in construction. In L. Clarke, E. Frydendal-Pedersen, E. Michielsens, B. Susman, & C. Wall (Eds.), *Women in construction* (pp. 158–172). Brussels: CLR.

Watts, J. H. (2007). Porn, pride and pessimism: Experiences of women working in professional construction roles. *Work, Employment & Society, 21*(2), 299–316.

Watts, J. H. (2009). 'Allowed into a man's world' meanings of work-life balance: Perspectives of women civil engineers as 'minority' workers in construction. *Gender, Work & Organization, 16*(1), 37–57.

Whittock, M. (2002). Women's experiences of non-traditional employment: Is gender equality in this area a possibility? *Construction Management & Economics, 20*(5), 449–456.

Wolkowitz, C. (2006). *Bodies at work*. London: Sage.

5

Change and Continuity: What Can Construction Tell Us About Institutional Theory?

Paul Chan

Introduction

In the bestseller turned Hollywood movie, *The Big Short*, Michael Lewis traced how a group of Wall Street traders spotted the then growing problem of subprime mortgages and capitalised on the housing bubble by shorting the market. By anticipating a drop in the value of the housing market, these traders made a fortune, but sounded a cautionary tale against the age-old belief that investments in real estate were as safe as houses. Despite the rupture to the global financial system brought about by the collapse of the housing market, little appears to have changed. The antidote to the financial crash was to inject more stimulation through quantitative easing and suppressing interest rates to maintain the buoyancy of the property market. In a review of *The Big Short*, a journalist stressed that despite the financial

P. Chan (✉)
School of Mechanical, Aerospace and Civil Engineering,
The University of Manchester, Manchester, UK
e-mail: Paul.chan@manchester.ac.uk

© The Author(s) 2018
D. J. Sage and C. Vitry (eds.), *Societies Under Construction*,
https://doi.org/10.1007/978-3-319-73996-0_5

151

crash of 2008, "people became hooked on profit, and continued believing housing could, and should, make money, rather than just provide shelter" (Foster 2016).

It has long been accepted that the construction industry serves as an important barometer for gauging the state of the economy. When Britain voted to leave the European Union in 2016, shockwaves were first felt by the property market. Similarly, the announcement of infrastructural building projects shortly after the election of Donald Trump as the 45th President of the USA also saw a rise in the Dow Jones Index. It would appear that the power of the construction industry in driving economic growth is unchanging and unchangeable. In this chapter, we question the seemingly stable conceptualisation of the construction industry by reviewing studies on organisational institutionalism in construction. This review highlighted how institutional scholarship, especially in the wider field of organisational and management studies, has shifted from a focus on stability and continuity, to a focus on flexibility and change. Early scholarship was influenced by concepts of institutionalisation and isomorphism (e.g. Meyer and Rowan 1977; DiMaggio and Powell 1983), as well as the institutional pillars of regulation, culture and norms (Scott 2001). Instead of viewing institutions as concrete things, more recent scholarship pays attention to how institutions are created, maintained and disrupted as they go through ongoing change through institutional work (Lawrence et al. 2011, 2013).

While change is fundamental to contemporary institutional scholarship, there is still a tendency to treat change as a transformation of static states. Such a view has been criticised for ignoring a processual reading of how organisations are changing (Weick and Quinn 1999; Chia 2002; Tsoukas and Chia 2002). In this chapter, I join this line of critical thinking to consider the implications a refocus on organisational changing can have on institutional scholarship, and propose ways in which the study of the construction industry can contribute. These proposals highlight three possible turns, including a turn towards a more processual and inclusive reading of institutional change, and a more productive reading of institutional demise.

Using Institutional Theory to Study Practices in the Construction Industry

Institutional Theory as Counterpoint to Formal, Rational Economic Models

One of the earliest studies to mobilise institutional theory to examine the construction industry was Oliver (1997). Through a survey of 296 residential building firms in Ontario, Canada, she differentiated between the effects of institutional relations and task environment on the firm's profitability. Prior to the framing of institutional theory (see Meyer and Rowan 1977; DiMaggio and Powell 1983), the success of a firm was often attributed to the rational, purposive actions of managers who were seen as controllers of organisational tasks and resources. Oliver's (1997) study highlighted the limits of such a view and noted how the quality of relationships between the firm and the institutional environment—regulatory agencies and professional associations—played a significant part in moderating a firm's success.

As Ball (1998) noted, interest in institutional analysis grew out of discontent with traditional economic models that are too rigid to explain the "messy real world of [...] cities and the buildings within them" and, "above all, the mysterious force of power [that] lurks within the property world" (p. 1501). He also added that while there is surprisingly an absence of a clear definition of what an "institution" is, "much of the institutionalist literature formally distinguishes between organisations (the players) and institutions (the rules)" (Ball 1998, p. 1502). It is the interplay between what the players or organisational actors do and the institutional rules that govern such actions and practices that have motivated much institutional analysis in construction. For example, in a study of leadership in construction, Bresnen (1995, p. 510) noted that "leadership is neither an individual fact nor a social fiction," as he explained how leadership is both "an abstracted myth that serves to legitimize and reinforce broader social norms and values," while at the same time "filtered, interpreted, and acted upon in very different ways" by practitioners.

Institutional Theory as a Way to Examine Variations

One of the most popular frameworks of institutional analysis deployed by students of construction is Scott's (2001) tripartite model of regulative, normative and cultural-cognitive elements. As Mahalingam and Levitt (2007, p. 523) explained in their analysis of conflicts in cross-cultural projects, the study of institutions which they defined as "a set of norms, rules and values operating in a given environment" can help researchers and practitioners understand how "regularity of behaviour among actors affected by that environment" is achieved. Instead of using the abstract notion of cultural difference to explain differences in practices of an American company operating in Germany and France, Mahalingam and Levitt (2007) used Scott's three institutional pillars to "unpack" more tangibly the rules, norms and values that help shape differing practices across the project life cycle. In a similar vein, Chi and Javernick-Will (2011) also used Scott's framework to consider how political culture and industrial structure account for differences in the arrangements of managing large-scale, high-speed rail projects in China and Taiwan.

Whereas these scholars examined differences in practices in global projects, others have used institutional analysis to analyse convergence in practices. For example, Low et al. (2009) investigated why companies would adopt sustainable practices through a case study of an Eco-city joint venture project between China and Singapore. Also drawing on Scott (2001), they suggested that companies would comply with calls for greener practices because more stringent environmental protection regulations would mean that non-compliance would result in higher costs, and compliance would increase their legitimacy given social norms and growing interest in valuing the environment (see also Gluch et al. 2009).

Institutional Theory, Legitimacy and Change

Staying with the environmental theme, Gluch (2009) studied how environmental professionals gain legitimacy in the construction industry and avoid being seen as "nitpicking nags"; she noted how such professionals needed to familiarise themselves with the 'rules of the game' and align with "the social patterns that guide agency, power and practice in a specific setting" (p. 966). In a similar vein, Hoffman and Henn (2008) cited an example of how a plumbing contractor may resist the idea of a more sustainable waterless urinal simply because the contractor stands to lose the contract to install copper pipes normally associated with a standard urinal. Drawing on Bechky (2006), they commented that, despite construction projects being configured as temporary organisations, such organisations revolve around "enduring, structured role systems" (Bechky 2006, p. 4), such that "when green is added to a standard construction project, the roles and relationships among the various actors become rearranged into a form that is outside the standard operating procedure [thus inviting] resistance" (Hoffman and Henn 2008, p. 400).

While conforming to established ways of working to gain legitimacy has been the focus of much institutional analysis in construction, institutional theory has also been used to explain innovation and change in the construction industry. For example, Kale and Arditi (2010) assessed innovation diffusion in the Turkish construction industry by considering DiMaggio and Powell's (1983) ideas on isomorphic change where institutional pressures can lead to the establishment of new practices through coercion, imitation or social norms. By examining the uptake of CAD technology and ISO 9000 in Turkish construction, Kale and Arditi's (2010) conclusions point to the power of imitative behaviour where "the fear of losing competitive advantage forces architectural design firms [...] to imitate other firms that have adopted innovation," so as "to send credible signals to their clients/customers that they follow state-of-the-art developments in their processes and operations" (p. 336).

The power of mimetic isomorphism in driving new practices in construction has also been acknowledged elsewhere. Dulaimi et al. (2003), for instance, surveyed clients, consultants and contractors in Singapore and found that the propensity to innovate depended on inter-organisational commitment across the supply chain, and specifically how firms perceive the level of interest by other firms to innovate. Similarly, Phua (2006) found in her survey of 526 firms in Hong Kong that the willingness to adopt partnering depended not on the rationale of efficiency, but on a firm's perception of institutional norms; the more a firm perceives partnering to be the norm in other firms, the more likely partnering will be adopted.

Institutional explanations have also been applied to answer the question as to why new, innovative practices do not break through established norms in the construction industry. Boland et al. (2008), for instance, commented on how "contracts and accepted practices on construction projects have evolved with the assumption that design and construction information will be represented with two-dimensional images and associated text documents," and how "standards have evolved in which the 2-D drawings are expected," which presuppose "that the sequencing of the design, bidding, and construction will be in discrete stages, with a full set of design documents put out for competitive bidding by contractors" (p. 905). Thus, practices become embedded and sedimented in a repertoire of physical artifacts, rules and standards that make it difficult for new practices (e.g. 3-D representations) to emerge and become accepted. A similar conclusion was reached by Harty and Whyte (2010, p. 471) who observed "a robust social division of labor [...] between 'designers' and 'drafters' [...] has persisted since before the advent of CAD technologies," as they argued that "it is, then perhaps no wonder that a move toward wholly digitally oriented methods of designing, drafting and collaborating was not entirely successful." There is therefore a deeply entrenched logic in the organising of construction work that drives resistance to these technological changes and innovation.

Institutional Logics in Construction

Nevertheless, such logic is not universal. A growing strand of institutional scholarship has, over the past decade, become more nuanced in the assessment of institutional logics at play in shaping organisational practices. Lounsbury (2007) argued for the shift away from "associated notions such as 'institutionalization' and 'isomorphism'" to conceptualise "institutional environments as more fragmented and contested [...] and influenced by multiple, competing logics" (p. 289). Drawing on Friedland and Alford (1991), Thornton and Ocasio (2008) explained that institutional logics are embedded in "the contradictory practices and beliefs inherent in the institutions of modern western societies," and that these institutions include "the capitalist market, the bureaucratic state, families, democracy and religion" (p. 101). Thornton and Ocasio (2008) added that "while institutions constrain action they also provide [...] individuals, groups and organizations with cultural resources for transforming individual identities, organizations, and society" (ibid.). As Seo and Creed (2002) noted, there is increasing evidence that contradictions are "a fundamental driving force for institutional change" (p. 226; see also Miettinen and Virkkunen 2005; Nicolini et al. 2016).

Yet, where the construction industry is concerned, studies seem to suggest that competing logics provide a fundamental source for institutional continuity and maintaining the status quo. For instance, in a longitudinal study of French acquisitions of UK companies, Mtar (2010) observed how the highly localised character of construction, unlike the more global outlook of manufacturing, meant that it was difficult to reproduce the French way of managing in the acquired British firm. Mtar (2010) noted that whereas the typical French project manager would be expected to oversee all technical aspects of managing a project, the technical competence of the British project manager was much lower because of the separation of professional roles and responsibilities in the design and construction process as a result of the British contracting system. She also recounted how one interviewee explained the peculiarity of the quantity surveyor in the British system serving the function of cost control yet it is the project manager who spends the money. Consequently, Mtar (2010) distinguished between the contradictory financial logic in the British construction industry and the technical logic in the French construction industry, and concluded that the incompatibility of the two dominant logics meant that the British-acquired firm was not able to reproduce the management practices of the French-acquiring firm.

In a study of the Dutch construction industry, Sminia (2011) examined how and why the longstanding practice of collusion among civil engineering contractors in bidding for projects persisted in 2001 despite the European Union declaring such a practice illegal in 1992. Sminia (2011) surmised that one of the contributing factors for such institutional continuity lay in the competing logics of price versus value when seeking out contractors in the tendering process. Because it is known that the commissioner of the construction projects decides on an offer based on the lowest price bid, contractors fear this would result in the erosion of their profits, reductions in the value of the project and ultimately threaten the viability of the business. As such, contractors view the emphasis on the lowest price tender to be unfair, thereby legitimising the continuation of their collusion activities, driving what is known as the "pre-consultation" practice underground following the declaration of its illegality in 1992.

In a review of why there is slow uptake of low-carbon technological innovations in the housing refurbishment market in the UK, Killip (2013) argued that the project-based nature of the construction industry meant that the conservativeness of the key players in the sector was borne out of an emphasis on the logic of "buildability", which in turn reduces the willingness to experiment. Consequently, the institutional framework is created and maintained to govern practices of new-build and not refurbishment. Thus, it is not simply a resistance to change, but "a reflection of the reality that systemic innovations take time to become established" (p. 882), signalling a need to ensure that those at the coalface are not alienated by innovation that is unfamiliar to them.

More recently, Oliveira et al. (2017) studied the institutional logics of how four international architectural practices embraced new practices of energy modelling. Through three overlapping logics of dependence, investment and risk, they sought to examine how the opportunities and constraints that energy modelling can bring in transforming architectural practice. Oliveira et al. (2017) found that while new technologies are emerging to facilitate early energy modelling in architectural design, the potential for these technologies to become game-changing is limited in part because designers and clients are still operating within existing rationalities, identities and expertise that do not pay attention to energy modelling. Their findings resonate with what some scholars highlighted as a more likely scenario of the emergence of hybrid practices between the old and the new (e.g. Boland et al. 2008; Harty and Whyte 2010). As Misangyi et al. (2008) stress, "the legitimacy of a new institutional logic depends on how well it ties into the existing logic, and because opponents to change will attempt to exploit any inconsistencies between the new and existing logics, the emergence of a hybrid institutional logic is highly probable" (p. 766).

Putting Institutional Theory to Work

A final strand of institutional scholarship that has gained traction in recent times can be found in the area of institutional work. In this line of thinking, the importance and role of self-aware, reflexive individuals are highlighted in their active engagement in the "processes of institutional creation, maintenance, disruption and change" (Lawrence et al. 2011, p. 53). Scholarship on institutional work sought to bring agency back into institutional analysis and to pay attention to what happens at the institutional coalface (Barley 2008). As Hardy and Maguire (2008) argued, organisational actors "are not viewed simply as carriers of institutional meanings associated with practices, diffusing them intact and unchanged through a field; rather, all actors in the field are viewed as active interpreters of practices whose meaning is, as a result, negotiated in ongoing, complex processes" (p. 205). It is these complex processes that the institutional work literature seeks to unravel.

The institutional work literature has paid quite substantial attention to the role and work done by professionals (see, e.g., Muzio et al. 2013; Bresnen 2016) to address what is known as the paradox of embedded agency (e.g. Friedland and Alford 1991; Seo and Creed 2002); that is, if individuals are subjected to the regulative, normative and cognitive processes such that their practices are influenced by the institutional field, how then do individuals come up with new practices and encourage others to adopt these? In a contemporary example of how environmental expertise is put to work in construction, Gluch and Bosch-Sijtsema (2016) studied the institutional work done by three environmental experts as they negotiated the multiple, everyday tensions between practice, agency and institution. They noted, for instance, how when project members did not adopt the new environmental practices, the environmental expert felt they were "lonely 'outsiders'" and had to downplay their expertise and knowledge "in order to avoid misunderstandings and create acceptance of new practices" (p. 532).

In another example, Daudigeos (2013) studied how professionals in occupational health and safety (OHS) gained legitimacy and facilitated the acceptance of safety practices in a VINCI, a large French construction company. Daudigeos (2013) noted how these OHS professionals were not always regarded as integral to the project team, and how they had to use their ability to network with, for instance, the sales team to promote the idea of and need for safer practices to the customer. He also described how OHS professionals sometimes repeated "rational myths" circulating among the project professions to frame the importance of the safety issue, gain acceptance from the project team and make up for their lack of formal authority in project delivery.

The institutional work literature not only focuses on what people do but also enriches our understanding of, as Jones and Massa (2013) put it, "the interplay of ideas, materials, and identities" (p. 1101). Schweber and Harty (2010), for instance, traced how reinforced concrete gained gradual acceptance as a novel building material. While not framed as an institutional analysis, Schweber and Harty's (2010) historical analysis highlighted the institutional work that helped drive the uptake of reinforced concrete in France while hindering its acceptance in Britain. They noted how, in France, François Hennebique's patented system coupled with his entrepreneurial spirit and strong networks allowed for rapid commercialisation of reinforced concrete in French building in the early 1900s. Schweber and Harty (2010) traced how Hennebique's enterprise led to a growing scientific culture around reinforced concrete and the professionalisation of reinforced concrete contractors. By contrast, in Britain, the Royal Institute of British Architects along with other professional organisations fought to break the monopoly of specialist firms over the use of reinforced concrete which delayed the professionalisation of these specialist firms in regulating and refining the use of reinforced concrete, with the consequence of falling standards in reinforced concrete buildings in the early twentieth century.

In another study, Jones and Massa (2013) described in great detail how the agency of Frank Lloyd Wright, among others, evangelises the novel use of reinforced concrete in the building of churches to the architectural profession. They recounted how Frank Lloyd Wright had to convince church members to break with the tradition of the ornate Gothic design so that a more efficient, standardised design that used reinforced concrete, a relatively mundane material when compared to the conventional stone used in church building could be delivered. Frank Lloyd Wright successfully argued that his design was more appropriate to beliefs of the members of the Unity Church as he persuaded the church members that "God should not be sought in the sky, but on earth among the children of men" (Jones and Massa 2013, p. 1111).

(Re-)Thinking Change Through an Institutional Lens: Possibilities Afforded by the Construction Industry

In a recent thought-piece, Bresnen (2017) was puzzled by the relative neglect of institutional theory in the construction management literature. Although the assertion that institutional theory makes an important contribution to organisational and management theory is reasonable, the preceding section shows—contrary to the claim of "virtually no papers informed by institutional theory" (Bresnen 2017, p. 25)—there have been quite a number of studies that mobilised institutional theory to examine the affairs and practices of the construction industry. What is striking, nevertheless, is the way institutional theory is used by scholars in the construction management field and wider field of organisational and management studies. As mentioned in the preceding section, scholars in the construction management field have tended to emphasise Scott's (2001) framework of regulative, normative and cognitive elements to examine how practices converge or diverge between different contexts. For these scholars (e.g. Mahalingam and Levitt 2007; Chi and Javernick-Will 2011), institutional theory is used to examine stabilisations through constructs of institutionalisation and isomorphism. By contrast, for scholars in wider field of organisational and management studies (e.g. Daudigeos 2013; Jones and Massa 2013), the emphasis has shifted from a focus on stabilisation to examining the dynamics of change by bringing in contemporary readings of institutional logics and institutional work.

Change, rather than continuity, has informed current understanding and theorising of organisational institutionalism. Yet, in the conceptualisation of change, there is a remarkable difference between scholars in the construction management field and those in wider organisational and management studies. For the former, the institutions that govern the practices in the construction industry—the rules and conventions of design and construction, the contracting system, the conservative nature of construction firms and in the UK, the disciplinary boundaries between the various professional groups—appear to be unchanging (and even unchangeable), often providing an explanation as to why the construction industry is lagging behind in terms of innovation. Commenting on the various industry reforms in the construction sector, for example, Fernie et al. (2006) argued that these change programmes are anything but, and that reforms towards continuously improving the construction industry are simply "reiterating familiar concerns" (p. 93; see also Green 2011). More recent reviews have begun to challenge this idea of a relatively stable, unchanging construction industry (e.g. Gottlieb and Jensen 2016).

Change is fundamental in any institutional analysis. In the remainder of this section, we first assess how change has been considered through the institutional theoretical lens. It is argued that change is largely treated as an entity and that the focus on states of change obscures a deeper, more processual understanding of change in general. Following this assessment, we conclude with a few suggestions as to what the construction industry can offer in terms of deepening our processual understanding of change through an institutional theoretical lens.

Change Management and Institutional Theory

Change is often treated as a relatively stable end point. Much less attention is paid on living through change (or changing to quote Weick and Quinn 1999). Organisational change is often conceptualised in relation to stability (McGuire and Hutchings 2006). So, in planned approaches to change, organisational change was seen as something to be tamed, often through managerial actions, in order to restore stability of organisation. Such conceptualisations responded either to exogenous (e.g. technological innovation, increased competition and intensification of globalisation) or endogenous (e.g. new managerial policies and organisational restructuring) forces of organisational change. Much early scholarship focussed on how change could be made acceptable to organisational members through effective communication and leadership so that survival and continuity of organisational routines could be secured (see Tichy 1983; Kotter 1996). Resistance to change and organisational inertia was seen as problematic to the production of effective responses to change (see Hannan and Freeman 1984; Argyris 1985). Yet, there are consistently high failure rates in organisational change management programmes (e.g. Dunphy and Stace 1988; Beer et al. 1990; Zairi et al. 1994; Kotter 1996; Orlikowski and Hofman 1997).

Scholars subscribing to planned approaches to change tend to privilege managerial mechanisms for addressing change, often downplaying the role of social agents and context (Orlikowski 1996). By contrast, scholars adopting a processual approach to researching organisational change took the view that change is an inevitable reality in organisations. What mattered more was the ability for researchers to re-present the socially constructed (Berger and Luckmann 1967) stories of the way change unfolds in practice over time, carefully attending to the importance of context and politics of change and situated actions (see Pfeffer and Salancik 1978; Suchman 1987; Burnes 1996; Dawson 1997). Processual researchers tend to adopt a more practice-based perspective of organisational change so as to seek explanations of how social agents make sense of ongoing change and consequent shaping of the acts of organising (see Weick 1995; Hendry 1996; Barley and

Tolbert 1997; Dawson 2003). Processual researchers of organisational change tend to avoid prescriptions for enforcing and reinforcing organisational conduct, and instead emphasise more on the constitution and institutionalisation of meanings and identities in organisations (see Greenwood et al. 2002). Stability, for processual researchers, is thus not the object but reified abstractions through the observations of researchers (Chia 2002). Put another way, stability in organisations can only be temporary; taken-for-granted practices in organisations can often be disrupted as a result of macro-level institutional pressures and thus subjected to continuous institutionalisation, deinstitutionalising and re-institutionalising forces (Greenwood and Hinings 1996).

As our preceding section has demonstrated, more recent scholarship on organisational institutionalism has been extremely productive in accounting for the dynamics of organisational change. Organisational institutional theorists were initially concerned with the diffusion of change through the dynamics of coercive, mimetic and normative isomorphic forces (Meyer and Rowan 1977; DiMaggio and Powell 1983), the translation of ideas (Czarniawska and Joerges 1996; Meyer 2008) and the articulation of institutional logics (Thornton et al. 2005; Thornton and Ocasio 2008) behind organisational change. Others have been puzzled by how new practices can be introduced into a context of established ones (Lounsbury 2007; Lounsbury and Crumley 2007), with growing interests in the role of institutional entrepreneurs (Maguire et al. 2004; Hardy and Maguire 2008) and institutional work (Lawrence et al. 2011, 2013) in mobilising change. Organisational routines have also been scrutinised (Feldman 2000; Feldman and Pentland 2003) to see how their performances are ostensibly linked with flexibility and change. Thus, the rich field of organisational institutionalism provides a more textured, fine-grained analysis of the dynamics of change processes to show not only convergence to effects of change observed, but also the contestations that lead to the institutionalisation of organisational change.

Despite the richness of change perspectives offered by organisational institutional theorists in the way change and stability are conceptualised, both planned and processual scholars share a common vantage point. That is, change is a phenomenon that is "out there" waiting to be analysed, rather than experienced "in here" (Pettigrew et al. 2001). Change, or rather the effects of change, is often the point of origin for many of the studies on organisational change. In Rao's et al. (2003) study of how nouvelle cuisine had displaced the classical style in French gastronomy, there was clear recognition of the change event/effect, i.e. the growing interest in nouvelle cuisine. Similarly, Lounsbury's and Crumley (2007) interest in tracing how new practices can arise in an institutional context began by noting the change—the rise of contracting—in governance structures of accounting firms. In examining the role of institutional entrepreneurs, Czarniawska (2009) revisited the historical context of the emergence of an institution, the London School of Economics and Political Science, to explain how ideas of its formation were translated through time to become the established institution it has become today. Vaara and Monin (2010) examined the role of discursive practices in the recursive relationship between legitimation and organisational action in distilling the dynamics of mergers and acquisitions. More recently, Hsu et al. (2012) analysed how the production of genre-spanning films has survived the penalties of hybridisation of organisational and institutional categories.

Similar patterns can be seen in the scholarship of institutional change in the construction industry. Phua (2006), for instance, surveyed companies to investigate the extent to which partnering as a new procurement practice was adopted. Others have studied what happens when computer-aided design is introduced, or when two-dimensional representations in drawings give way to three-dimensional models (e.g. Boland et al. 2008; Kale and Arditi 2010; Harty and Whyte 2010). Gluch and colleagues also studied how environmental expertise becomes an acceptable norm in the construction workplace (Gluch 2009; Gluch et al. 2009; Gluch and Bosch-Sijtsema 2016). In many accounts of organisational change within the construction management literature and wider organisational and management studies, change is usually concretised as an end point in sight for the researchers and practitioners to make sense of Rao et al. (2003), for instance, set the year 1997 as the end of the observational period "because by then, *'cuisine sous contrat'* and *'cuisine rassurante'* (comforting and reassuring cuisine) had arisen on the scene" (p. 798). In other words, a new change is in sight.

There is growing dissatisfaction with the ontological and epistemo-logical standpoints occupied in the production of knowledge about organisational change (Van de Ven and Poole 2005). Pettigrew et al. (2001) acknowledged the challenges of understanding the dynamics of organisations and claimed that much research on organisational change has, in the main, accounted for static states. Despite the fine-grained analyses provided by organisational institutional theorists, change is invariably reduced to episodic moments of deinstitutionalisation, insti-tutionalisation and re-institutionalisation. In a sense, such episodic moments of understanding change still bear semblance to the unfreez-ing-moving-refreezing thesis offered by Lewin (1947).

Scholars have raised concerns over the way change is studied. Weick and Quinn (1999) argued that the focus should be diverted to ongoing change, or changing, in organisations. Chia (2002), inspired by process philosopher Henri Bergson's (see Bergson 2008 [1910]) thinking about time, argued against the way knowledge about change has become a "com-modified 'product'; a thing that can be 'assembled', 'rearranged', 'pack-aged', 'transferred' and 'consumed'" (p. 865). Of contention is the way change and change processes are deciphered and made distinguishable. Change can thus not only be understood through its visible effects over time, but also by the everyday lived experiences of all those concerned. According to Barley (2008), organisational institutional theorists have largely been concerned with linking macrosocial theory with micro-social practices, while ignoring "the throws of everyday life" (p. 510).

There is, therefore, a need to understand change through everyday changing where the action is "in here" (Pettigrew et al. 2001, p. 700) with all its incompleteness, in-between-ness and ambiguities (Ellis and Ybema 2010; Beech 2011), and where the end point of change is not clearly in sight. As Chia (2002) argued, the process of change must be viewed as "ceaselessly becoming" (i.e. a movement) rather than "assumed to be pro-cesses of primary 'things'" (p. 866, original emphasis). He stressed that to deepen our understanding of the movement of change one has to suspend the "tendency to think about change as change of 'something' and to argue about movement as though it is the movement of an entity" (Chia 2002, p. 865). Thus, change is "endogenously conditioned, and it cannot be fully anticipated" (Tsoukas and Chia 2002, p. 578).

Possibilities Afforded by the Construction Industry in Examining Institutional Change

How does the foregoing discussion on organisational change and changing add to general understanding of institutional theory? What possibilities are afforded by the construction industry in studying change and changing through an institutional lens? In this section, we sketch out three possibilities that could open up productive lines of inquiry into change and changing using an institutional theoretical lens. These include the possibility of paying more attention to the microfoundational processes of change, the possibility of more inclusive scholarship and practice, and the possibility of demise.

Towards More Processual Accounts of Institutional Change

Institutional theories have accounted for the organisational tensions between stability and change. Early theorists have emphasised how institutions can constrain what organisational actors do and bring relationships in and around organisations to order, while later theorists focus on change and how new practices and institutions can actively be created and maintained. This shift has seen institutional scholars move from an entitative framing where institutions are concrete things that structure organisational actions and behaviour, to a processual framing where institutions are constantly changing. This reframing of institutions as an entity to institutions as the process has also seen a methodological transformation, from an early focus in searching for a variance to contemporary interest in capturing the dynamics of change through qualitative, longitudinal and increasingly ethnographic studies (see Surachaikulwattana and Phillips 2017). As Whittington (2017) noted, a processual framing complements an institutional account of organisations for "a Process view increases the temporal range of activity; cross-sectionally, Institutionalism captures activity as expressed in society-wide practices" (p. 388).

A processual framing of institutions has brought agency to the fore as evident in the growing scholarship on institutional entrepreneurship and institutional work (see Maguire et al. 2004; Barley 2008; Hardy and Maguire 2008; Maguire and Hardy 2009; Lawrence et al. 2011, 2013; Daudigeos 2013; Gluch and Bosch-Sitjsema 2016). While this body of work enriches our understanding of how institutions and organisational practices are changing, there have been over the past decade growing criticism of the (false) dichotomy between stability and change. In taking a processual viewpoint, organisational scholars have often seen the world not in terms of nouns but verbs; adhering to a strong process ontological worldview, everything is seen as a reflexive process of sense-making where organisations are in constant flux of performing activity (see Chia 2002; Tsoukas and Chia 2002; Langley et al. 2013; Mutch 2016; Whittington 2017). Yet, as Bakken and Hernes (2006) argued, drawing on the complementarity between Karl Weick and process philosopher Alfred North Whitehead, organising requires both nouns and verbs, and it is important to acknowledge the "*relationality* between verb and noun, where verbs are made from nouns, and vice versa. Rather than being seen as ontologically different from verbs, nouns may be seen as temporarily stabilized configurations of recurring processes, which are given labels" (p. 1612, original emphasis). Thus, it is crucial that we remember the gerund form of the word organising, where the verb form functions simultaneously as a noun. Instead of emphasising the antagonistic dualism of stability and change, Farjoun (2010) argued for considering the paradoxical duality of stability and change, where "stability and change are not separable and only conflicting, but, rather, they are fundamentally interdependent" (p. 216). He added that seeing stability and change as both sides of the same coin would mean that the "dominant view of organizations as rigid and myopic or more mindful and inventive" (ibid.) would be dismissed.

A productive and fertile area of organisational scholarship that has acknowledged this duality of stability and change can be found in recent studies of organisational routines. Feldman (2000) and Feldman and Pentland (2003) first drew our attention to organisational routines not as an automatic repetition, but as effortful accomplishments shaped by both performative and ostensive aspects. Put another way, no routine is ever repeated in the same way twice. For Feldman and colleagues, routines are simultaneously (and not sequentially) a source of stability as they are a source of change; every routine, they maintained, requires both the act of doing and the act of patterning. It is this act of doing (the performative) and the act of patterning (the ostensive) that give rise to what Birnholtz et al. (2007) called the paradox of the (n)ever-changing world.

Organisational scholars have grappled with the challenge of unravelling this paradoxical character of routines. For instance, Pentland et al. (2011) studied invoicing routines in four organisations (including a construction company) and observed how even though "the entry and approval routines in these four organizations involved the same technology, same lexicon of action, and same goals […] these routines generated patterns of action that changed over time" (p. 1379). More recently, Dittrich et al. (2016), through a year-long ethnographic study, found that reflective talk played an important role in driving routine change in a pharmaceutical start-up company. Specifically, Dittrich et al. (2016) observed how reflective talk allowed organisational actors to envisage and explore variations in situated actions in a specific performance while simultaneously abstracting the pattern of the routine; thus, reflective talk "is an important means by which actors produce variations in a routine and select some of these for retention" (p. 692). This recursive and discursive character of routine change also featured, to some extent, in Hopkin's et al. (2016) study of how housing associations learn from defects in new-build housing projects.

The paradox of a (n)ever-changing world, where stability and change are inseparable, represents a potential line of scholarship in institutional change. Rather than to view these contradictions as essentially opposing and problematic, organisational scholars interested in the processual dynamics of institutional change should embrace (and not eradicate) contradictions as a powerful resource for propelling organisational actions and behaviour (Seo and Creed 2002; Miettinen and Virkkunen 2005; Nicolini et al. 2016). It is here that the construction industry offers a fertile empirical basis for studying paradoxes. Szentes (2016), for instance, noted how the project-based nature of the construction industry meant that organisational actors often have to balance between exploration and exploitation, between short-term demands and long-term survival, and between the need for standardisation and control on the one hand and the need to flexibility and change on the other (see also Szentes and Eriksson 2016; Jia et al. 2017). Robinson et al. (2016a, b) drew inspiration from Activity Theory (see Engeström 2000; Miettinen and Virkkunen 2005) and the concept of affordances (Leonardi 2011) to study how a construction contractor made sense of contradictions that emerged as they moved away from a product-centric towards a more service-centric way of working. Their conclusions highlight a rich array of specific technological components and features (e.g. sensors and spreadsheets) in construction that present opportunities for studying how routines stabilise and shift over time through the interplay of ideas, materials and identities (see also Jones and Massa 2013). In so doing, Robinson et al. (2016b) provides an example of departing from existing retrospective case study research and correcting what Lawrence et al. (2013) observed as an overlooked aspect in current scholarship in institutional work; that is, the limits of "our ability to uncover and understand the messy day-to-day practices [… through] the mundane, ordinary ways in which institutions are embodied at a micro level and how actors engage with them in their day-to-day activities" (p. 1029).

Towards More Inclusive Accounts of Institutional Change

A processual approach to studying institutional change would, as Langley et al. (2013) suggest, allow the possibility "to study people, subject matter, and their context in meaningful ways" (p. 6), although this would require researchers to develop "their interactional expertise […] that engage [research participants] in sharing what they know, its technical content, and what is going on in the setting. This involvement provides researchers not only access to, but also an appreciation of, specialists' views, activities, and interests" (Langley et al. 2013, p. 6). This appreciation can be seen in Daudigeos' (2013) study of OSH professionals in construction, where he reflected on how "we were somewhat overwhelmed by technical considerations regarding OSH professionals' activities [… and] challenged on the variability of the informants' answers" (p. 730), admitting how "we were still in a learning phase" (ibid.). A processual approach to studying institutions and institutional change would therefore require researchers to develop their intuition in the field and to recognise the interactional and informal aspects of expertise (Chan 2016).

As outlined above, there are a growing number of studies that seek to capture how individual specialists undertake institutional work to create, promote, maintain and disrupt organisational practices. These tended to focus on professionals, such as the specialist reinforced concrete contractors in Schweber and Harty (2010), the OSH professionals in a French construction company in Daudigeos (2013), the environmental specialists in Swedish construction in Gluch and Bosch-Sijtsema (2016) and architects and building services engineers in Oliveira et al. (2017). Although these studies sought to reclaim the agency of individuals in institutional work, the focus has mainly been on organisational and professional elites. It is worth noting that, contrary to the literature on institutional entrepreneurs (e.g. Maguire et al. 2004; Hardy and Maguire 2008; Maguire and Hardy 2009), scholars in the institutional work stream have attempted to avoid falling into the trap of framing the institution worker as heroic (Battilana et al. 2009; Muzio et al. 2013).

By focussing on the institutional work done by the professional elites, there appears to be a relative, if unintended, neglect of the accounts of institutional work done at the coalface (Barley 2008). Indeed, it is difficult to find a study that focussed mainly on the accounts of workers and how they contribute to institutional change. In reviewing *Heathrow Terminal 5: History in the making*, Chan (2012) criticised the overwhelming focus on managerial actions and the remarkable absence of any account of the workers' contribution to the making of this monumental project. As Ness and Green (2013) argued, accounts of change in the construction industry have often neglected what they see as the "disappearing worker." Even though Ness and Green (2013) recognised how the discourse around improving the construction industry can shape the everyday realities for workers, their analysis falls short in showing how workers can resist such discourse and partake in institutional work.

Nevertheless, this presents another opportunity for expanding our narratives of institutional work through construction work. Just as Maguire and Hardy (2009) studied the institutional work entailed in delegitimising DDT, there is scope to undertake similar analyses of hazardous substances used in construction. A good example can be found in the use of asbestos as a building material. Calvert's (2010) analysis, for instance, highlighted the role trade unions could play in vocalising concerns about the continued endorsement of asbestos in the Canadian construction industry. In his reflection, he noted how conflicting responses between trade unions representing building workers and the wider labour federation stymied progress in banning the use of asbestos. Unlike Maguire and Hardy's (2009) study of the role outsiders play in delegitimising the use of the hazardous substance, there is potential for Calvert's (2010) study to consider the role insiders, and especially construction workers, play in institutional work in this arena. Another contemporary example can be found in the blacklisting of construction workers scandal in the UK, where workers were deliberately excluded from the construction labour market as a result of their trade union activities. While Druker's (2016) recent article was not framed as an institutional piece, her analysis highlights the possible pathways (and institutional work) that construction employers could take in either entrenching or challenging such anti-union practices as blacklisting of workers.

Towards More Productive Accounts of Demise in Institutional Change

The institutional literature is rich with studies that show how institutions are created, maintained and sometimes disrupted. However, there are relatively fewer studies that provide accounts of the erosion of institutions and the processes of deinstitutionalisation. Maguire and Hardy's (2009) study of the delegitimising of DDT is a notable exception, as is the study of institutional replacement of the global energy system from a traditional state-owned and state-run model to a neoliberal model of energy provision found in Henisz et al. (2005). Many institutional accounts of change, certainly in the construction management field, tend to focus on ushering in a brave new world, including studies of the novel use of reinforced concrete (Schweber and Harty 2010; Jones and Massa 2013), how computer-aided design and digital technologies that allow for representations beyond two-dimensional modelling can transform the design and construction process (Boland et al. 2008; Harty and Whyte 2010), and how environmental expertise can make the construction process more sustainable (Gluch et al. 2009; Gluch and Bosch-Sijtsema 2016). Even where a hybrid organisation that mixes the old and new logics of working is considered, particular emphasis is firmly placed on moving towards a brave new world of working. The "traditional" ways of working are often treated as a problem, but rarely problematised.

With every new practice comes a point of departure, an old practice ruined. More recently, de Cock and O'Doherty (2017) invited us to pay more attention to move beyond studying organisations as usual to consider the vitality of ruin in organisations. They argued that,

> The ruin lies like a hidden blueprint behind the ostensible reality of buildings and walls, offices, furniture, organizational charts, mission statement and all the other accoutrements of rational organization, but fragments of this becoming-ruin are evident if one can learn to look out of the corner of one's eye. The flecks of paint peeling from the walls, the carpet worn under the pressure of human traffic, the fading photographs on the wall, are all imprints of a time in organization that anticipates ruin writ large. (p. 143)

There is again much that the construction industry can offer in providing a fertile terrain for producing accounts of demise in institutional change. Buildings often outlive the purpose they were originally intended for. Warehouses and cotton mills that saw the heyday of the industrial revolution have become converted into retail outlets, restaurants and residential apartments. Innovative entrepreneurs in fab labs have now replaced regiments of blue-collar workers in factories of the past. In some cases, celebrated edifices like Olympic parks have given way to decay and decline. While much research in construction management focusses on new build, with a growing body of scholarship that attends to the importance of maintenance and renewal, there is still much to be discovered from the demise and demolition of buildings.

A recent example can be found in Grenfell Tower, a residential block of flats in an affluent part of London, which was destroyed in a fire in June 2017 resulting in the death of at least eighty people. At the time of writing this chapter, the inquiry into the Grenfell Tower tragedy was still at an early stage. However, questions have already thrown into the spotlight the role of institutional arrangements in creating the conditions for the demise of Grenfell Tower. Residents of Grenfell Tower live in one of the UK's most economically deprived areas within a borough that is also one of the richest in the country. It would also appear that residents had previously raised concerns over fire safety and that cheaper, less fire-resistant external cladding was used in a recent refurbishment of the tower. The demise of Grenfell Tower has not only become a catalyst for rethinking the robustness of building and fire regulations, but also for reflecting on inequalities in society by questioning the adequacy of various government policies associated with the built environment, from housing and social policy to economic austerity.

The study of institutional change and ruin can enable us to examine how past institutional logics in the built infrastructure can provide clues of how history has shaped and can shape organisational and societal practices of the present. In a recent study, for instance, de Vaujany and Vaast (2014) examined the buildings and office spaces of Paris Dauphine University to figure out how remnants of their past life as NATO offices are made durable through inscriptions in the walls. They suggested paying more attention to what they called "spatial legacies", which "[encapsulate] the idea that, at any point in time, an organizational space displays traces of previous periods' spaces and spatial practices in ways that constrain and enable current spatial practices, the organizational space itself, as well as claims of organizational legitimacy. Over time, their material function may become obsolete, but they may keep a symbolic function" (p. 725). It is worth noting the multiple sources of evidence used in de Vaujany and Vaast (2014), which comprised participant observations, interviews and an array of archival visual and textual data to access the intentions of the original designers of the NATO building in the 1950s. With advances in building information modelling where asset management data can in principle trace over a lifetime the changing intentions of those who design, build and operate buildings, these information models could provide useful data sources for researchers attempting to analyse the impacts of the spatial legacies of the built environment and how these are transformed—made durable and decay—over time, and with what consequences for organisational and societal practices.

A final and perhaps more extreme form of demise that is especially peculiar in the construction industry is the notion of the workplace. It is well acknowledged that, being project-based and with construction sites being geographically localised, the construction industry relies heavily on a transient and in many cases peripatetic workforce. Yet, what constitutes the construction workplace also needs to be rendered problematic. Unlike most workplaces, the construction workplace goes through ongoing transformation, both socially and materially, as construction work proceeds. More crucially, taking the construction site in focus, the workplace also disappears as soon as work finishes. The implications of this inevitable demise of the workplace on site have rarely been examined. In accounting for institutional change, there is much scope for developing more productive narratives of demise through the example of the construction industry to see how processes of deinstitutionalisation, as much as of institutionalisation, can contribute to ever-changing organisational and institutional practices.

Concluding Remarks

At the time of writing, the world is undergoing institutional change. The resurgence in popular interest in the nation state threatens to dismantle decades of international cooperation through supranational institutions. The electorate in many countries have been seduced to the idea of regaining control over its borders, with commentators suggesting a social movement that rallies against the political establishment and societal elites. The built environment and contribution of the construction industry is restated in several ways in the unfolding narratives of these developments, from rhetorical calls to build walls along borders of countries, to the continued deindustrialisation of the developed world symbolised by the decline of industrial buildings, to the promises of infrastructure development and the building of affordable and safe homes. As the well-rehearsed argument goes, the construction industry underpins everything else that we do at work, in organisations and industry, and in society.

This chapter began with a portrayal of construction and the built environment as a seemingly unchanging and unchangeable institution. Of course, this depiction is merely impressionistic. In this chapter, how thinking about institutions has transformed over time has been outlined, from emphasising stability and continuity, to current conceptualisations of flexibility and change. The endeavour in this chapter was twofold. On the one hand, students of construction are encouraged to pay more attention to the interdependent aspects of stability and change, and to break free from ossified notions of institutions as illustrated through concepts of isomorphism, institutionalisation and institutional pillars. There is much scope to centre our attention on the paradox of stability and change, or what is known as the paradox of a (n)ever-changing world, so that we can examine how contradictions can propel activity. On the other hand, institutional theorists are encouraged to place more attention on the possibilities that the construction industry can offer in unravelling the processual dynamics of institutional change, so that we can have a more inclusive account of ways of changing, whether this means institutional change towards a brave new world or the study of demise in replacing established institutions.

References

Argyris, C. (1985). *Strategy, change and defensive routines*. Boston: Harvard University Press.

Bakken, T., & Hernes, T. (2006). Organizing is both a verb and a noun: Weick meets Whitehead. *Organization Studies, 27*(11), 1599–1616.

Ball, M. (1998). Institutions in British property research: A review. *Urban Studies, 35*(9), 1501–1517.

Barley, S. R. (2008). Coalface institutionalism. In R. Greenwood, C. Oliver, K. Sahlin, & R. Suddaby (Eds.), *The SAGE handbook of organizational institutionalism* (pp. 491–518). London: Sage.

Barley, S. R., & Tolbert, P. S. (1997). Institutionalization and structuration: Studying the links between action and institution. *Organization Studies, 18*(1), 93–117.

Battilana, J., Leca, B., & Boxenbaum, E. (2009). How actors change institutions: Towards a theory of institutional entrepreneurship. *Academy of Management Annals, 3*(1), 65–107.

Bechky, B. A. (2006). Gaffers, gofers, and grips: Role-based coordination in temporary organizations. *Organization Science, 17*(10), 3–21.

Beech, N. (2011). Liminality and the practices of identity reconstruction. *Human Relations, 64*(2), 285–302.

Beer, M., Eisenstat, R. A., & Spector, B. (1990). Why change programs don't produce change. *Harvard Business Review, 68*(6), 158–166.

Berger, P. L., & Luckmann, T. (1967). *The social construction of reality*. New York: Doubleday.

Bergson, H. (2008 [1910]). *Time and free will: An essay on the immediate data of consciousness*. New York: Cosimo.

Birnholtz, J. P., Cohen, M. D., & Hoch, S. V. (2007). Organizational character: On the regeneration of Camp Poplar Grove. *Organization Science, 18*(2), 315–332.

Boland, R. J., Sharma, A. K., & Afonso, P. S. (2008). Designing management control in hybrid organizations: The role of path creation and morphogenesis. *Accounting, Organizations and Society, 33*(7–8), 899–914.

Bresnen, M. J. (1995). All things to all people? Perceptions, attributions, and constructions of leadership. *Leadership Quarterly, 6*(4), 495–513.

Bresnen, M. (2016). Institutional development, divergence and change in the discipline of project management. *International Journal of Project Management, 34*(2), 328–338.

Bresnen, M. (2017). Being careful what we wish for? Challenges and opportunities afforded through engagement with business and management research. *Construction Management and Economics, 35*(1–2), 24–34.

Burnes, B. (1996). No such thing as… a "one best way" to manage organizational change. *Management Decision, 34*(10), 11–18.

Calvert, J. (2010). Canada's asbestos policy: An ongoing threat to building workers' health in Canada and around the globe. *CLR News*, Number 4-2010, 7–25. Retrieved from http://www.clr-news.org/CLR/Calvert%20final.pdf.

Chan, P. (2012). Review of the book *Heathrow's Terminal 5: History in the making*, by Doherty, S. *Construction Management and Economics, 30*(6), 498–500.

Chan, P. (2016). Expert knowledge in the making: Using a processual lens to examine expertise in construction. *Construction Management and Economics, 34*(7–8), 471–483.

Chi, C. S. F., & Javernick-Will, A. N. (2011). Institutional effects on project arrangement: High-speed rail projects in China and Taiwan. *Construction Management and Economics, 29*(6), 595–611.

Chia, R. (2002). Essai: Time, duration and simultaneity: Rethinking process and change in organizational analysis. *Organization Studies, 23*(6), 863–868.

Czarniawska, B. (2009). Emerging institutions: Pyramids or anthills? *Organization Studies, 30*(4), 423–441.

Czarniawska, B., & Joerges, B. (1996). Travel of ideas. In B. Czarniawska & G. Sevón (Eds.), *Translating organizational change* (pp. 13–48). Berlin: De Gruyter.

Daudigeos, T. (2013). In their profession's service: How staff professionals exert influence in their organization. *Journal of Management Studies, 50*(5), 722–749.

Dawson, P. (1997). In at the deep end: Conducting processual research on organisational change. *Scandinavian Journal of Management, 13*(4), 389–405.

Dawson, P. (2003). *Reshaping change: A processual perspective.* London: Routledge.

De Cock, C., & O'Doherty, D. (2017). Ruin and organization studies. *Organization Studies, 38*(1), 129–150.

de Vaujany, F., & Vaast, E. (2014). If these walls could talk: The mutual construction of organizational space and legitimacy. *Organization Science, 25*(3), 713–731.

DiMaggio, P. J., & Powell, W. W. (1983). The iron cage revisited: Institutional isomorphism and collective rationality in organizational fields. *American Sociological Review, 48*(2), 147–160.

Dittrich, K., Guérard, S., & Seidl, D. (2016). Talking about routines: The role of reflective talk in routine change. *Organization Science, 27*(3), 678–697.

Druker, J. (2016). Blacklisting and its legacy in the UK construction industry: Employment relations in the aftermath of exposure of the Consulting Association. *Industrial Relations Journal, 47*(3), 220–237.

Dulaimi, M. F., Ling, F. Y. Y., & Bajracharya, A. (2003). Organizational motivation and inter-organizational interaction in construction innovation in Singapore. *Construction Management and Economics, 21*(3), 307–318.

Dunphy, D. C., & Stace, D. A. (1988). Transformational and coercive strategies for planned organizational change: Beyond the organization development (O.D.) model. *Organization Studies, 9*(3), 317–334.

Ellis, N., & Ybema, S. (2010). Marketing identities: Shifting circles of identification in inter-organizational relationships. *Organization Studies, 31*(3), 279–305.

Engeström, Y. (2000). Activity theory as a framework for analysing and redesigning work. *Ergonomics, 43*(7), 960–974.

Farjoun, M. (2010). Beyond dualism: Stability and change as a duality. *Academy of Management Review, 35*(2), 202–225.

Feldman, M. S. (2000). Organizational routines as a source of continuous change. *Organization Science, 11*(6), 611–629.

Feldman, M. S., & Pentland, B. T. (2003). Reconceptualizing organizational routines as a source of flexibility and change. *Administrative Science Quarterly, 48*(1), 94–118.

Fernie, S., Leiringer, R., & Thorpe, T. (2006). Change in construction: A critical perspective. *Building Research and Information, 34*(2), 91–103.

Foster, D. (2016, January 29). Housing blew up the global economy in 2008 and we learned nothing. *Guardian.* Retrieved from www.guardian.co.uk.

Friedland, R., & Alford, R. R. (1991). Bringing society back in: Symbols, practices, and institutional contradictions. In W. W. Powell & P. J. DiMaggio (Eds.), *The new institutionalism in organizational analysis.* Chicago: University of Chicago Press.

Gluch, P. (2009). Unfolding roles and identities of professionals in construction projects: Exploring the informality of practices. *Construction Management and Economics, 27*(10), 959–968.

Gluch, P., & Bosch-Sijtsema, P. (2016). Conceptualizing environmental expertise through the lens of institutional work. *Construction Management and Economics, 34*(7–8), 522–535.

Gluch, P., Gustafsson, M., & Thuvander, L. (2009). An absorptive capacity model for green innovation and performance in the construction industry. *Construction Management and Economics, 27*(5), 451–464.

Gottlieb, S. C., & Jensen, J. S. (2016). Governmentalities of construction: From mortar to modular systems and markets. In P. W. Chan & C. J. Neilson (Eds.), *Proceedings 32nd Annual ARCOM Conference,* 5–7 September, Manchester, UK (pp. 3–12). Association of Researchers in Construction Management.

Green, S. (2011). *Making sense of construction improvement.* Oxford: Wiley-Blackwell.

Greenwood, R., & Hinings, C. R. (1996). Understanding radical organizational change: Bringing together the old and the new institionalism. *Academy of Management Review, 21*(4), 1022–1054.

Greenwood, R., Suddaby, R., & Hinings, C. R. (2002). Theorizing change: The role of professional associations in the transformation of institutionalized fields. *Academy of Management Journal, 45*(1), 58–80.

Hannan, M. T., & Freeman, J. (1984). Structural inertia and organizational change. *American Sociological Review, 49*(2), 149–164.

Hardy, C., & Maguire, S. (2008). Institutional entrepreneurship. In R. Greenwood, C. Oliver, K. Sahlin, & R. Suddaby (Eds.), *The SAGE handbook of organizational institutionalism* (pp. 198–217). London: Sage.

Harty, C., & Whyte, J. (2010). Emerging hybrid practices in construction design work: Role of mixed media. *Journal of Construction Engineering and Management, 136*(4), 468–476.

Hendry, C. (1996). Understanding and creating whole organizational change through learning theory. *Human Relations, 49*(5), 621–641.

Henisz, W. J., Holburn, G. L. F., & Zelner, B. A. (2005). *Deinstitutionalization and institutional replacement: State-centred and neo-liberal models in the global electricity supply industry* (Working Paper).

Hoffman, A. J., & Henn, R. (2008). Overcoming the social and psychological barriers to green building. *Organization and Environment, 21*(4), 390–419.

Hopkin, T., Lu, S.-L., Rogers, P., & Sexton, M. (2016). Detecting defects in the UK new-build housing sector: A learning perspective. *Construction Management and Economics, 34*(1), 35–45.

Hsu, G., Negro, G., & Perretti, F. (2012). Hybrids in Hollywood: A study of the production and performance of genre-spanning films. *Industrial and Corporate Change, 21*(6), 1427–1450.

Jia, A. Y., Rowlinson, S., Loosemore, M., Xu, M., Li, B., & Gibb, A. (2017). Institutions and institutional logics in construction safety management: The case of climatic heat stress. *Construction Management and Economics, 35*(6), 338–367.

Jones, C., & Massa, F. G. (2013). From novel practice to consecrated exemplar: Unity temple as a case of institutional evangelizing. *Organization Studies, 34*(8), 1099–1136.

Kale, S., & Arditi, D. (2010). Innovation diffusion modelling in the construction industry. *Journal of Construction Engineering and Management, 136*(3), 329–340.

Killip, G. (2013). Transition management using a market transformation approach: Lessons for theory, research, and practice from the case of low carbon housing refurbishment in the UK. *Environment and Planning C: Government and Policy, 31*(5), 876–892.

Kotter, J. (1996). *Leading change*. Boston: Harvard Business School Press.

Langley, A., Smallman, C., Tsoukas, H., & Van de Ven, A. H. (2013). Process studies of change in organization and management: Unveiling temporality, activity, and flow. *Academy of Management Journal, 56*(1), 1–13.

Lawrence, T. B., Leca, B., & Zilber, T. B. (2013). Institutional work: Current research, new directions and overlooked issues. *Organization Studies, 34*(8), 1023–1033.

Lawrence, T., Suddaby, R., & Leca, B. (2011). Institutional work: Refocusing institutional studies of organization. *Journal of Management Inquiry, 20*(1), 52–58.

Leonardi, P. M. (2011). When flexible routines meet flexible technologies: Affordance, constraint, and the imbrication of human and material agencies. *MIS Quarterly, 35*(1), 147–167.

Lewin, K. (1947). Frontiers in group dynamics. In D. Cartwright (Ed.), *Field theory in social science*. London: Social Science Paperbacks.

Lounsbury, M. (2007). A tale of two cities: Competing logics and practice variation in the professionalizing of mutual funds. *Academy of Management Journal, 50*(2), 289–307.

Lounsbury, M., & Crumley, E. T. (2007). New practice creation: An institutional perspective on innovation. *Organization Studies, 28*(7), 993–1012.

Low, S. P., Low, J. Y., & Liu, P. W. (2009). Sustainable facilities: Institutional compliance and the Sino-Singapore Tianjin Eco-city. *Facilities, 27*(9/10), 368–386.

Maguire, S., & Hardy, C. (2009). Discourse and deinstitutionalization: The decline of DDT. *Academy of Management Journal, 52*(1), 148–178.

Maguire, S., Hardy, C., & Lawrence, T. B. (2004). Institutional entrepreneurship in emerging fields: HIV/AIDS treatment advocacy in Canada. *Academy of Management Journal, 47*(5), 657–679.

Mahalingam, A., & Levitt, R. E. (2007). Institutional theory as a framework for analysing conflicts on global projects. *Journal of Construction Engineering and Management, 133*(7), 517–528.

McGuire, D., & Hutchings, K. (2006). A Machiavellian analysis of organisational change. *Journal of Organizational Change Management, 19*(2), 192–209.

Meyer, R. E. (2008). New sociology of knowledge: Historical legacy and contributions to current debates in institutional research. In R. Greenwood, C. Oliver, K. Sahlin, & R. Suddaby (Eds.), *The SAGE handbook of organizational institutionalism* (pp. 519–538). London: Sage.

Meyer, J. W., & Rowan, B. (1977). Institutionalized organizations: Formal structure as myth and ceremony. *American Journal of Sociology, 83*(2), 340–363.

Miettinen, R., & Virkkunen, J. (2005). Epistemic objects, artefacts and organizational change. *Organization, 12*(3), 437–456.

Misangyi, V. F., Weaver, G. R., & Elms, H. (2008). Ending corruption: The interplay among institutional logics, resources, and institutional entrepreneurs. *Academy of Management Review, 33*(3), 750–770.

Mtar, M. (2010). Institution, industry and power effects on integration in cross-border acquisitions. *Organization Studies, 31*(8), 1099–1127.

Mutch, A. (2016). The limits of process: On (re)reading Henri Bergson. *Organization, 23*(6), 825–839.

Muzio, D., Brock, D. M., & Suddaby, R. (2013). Professions and institutional change: Towards an institutionalist sociology of the professions. *Journal of Management Studies, 50*(5), 699–721.

Ness, K., & Green, S. (2013). Human resource management in the construction context: Disappearing workers in the UK. In A. Dainty & M. Loosemore (Eds.), *Human resource management in construction: Critical perspectives* (2nd ed., pp. 18–50). Abingdon: Routledge.

Nicolini, D., Delmestri, G., Goodrick, E., Reay, T., Lindberg, K., & Adolfsson, P. (2016). Look what's back! institutional complexity, reversibility and the knotting of logics. *British Journal of Management, 27*(2), 228–248.

Oliver, C. (1997). The influence of institutional and task environment relationships on organizational performance: The Canadian construction industry. *Journal of Management Studies, 34*(1), 99–124.

Oliveira, S., Marco, E., Gething, B., & Organ, S. (2017). Evolutionary, not revolutionary: Logics of early design energy modelling adoption in UK architecture practice. *Architectural Engineering and Design Management, 13*(3), 168–184.

Orlikowski, W. J. (1996). Improvising organizational transformation over time: A situated change perspective. *Information Systems Research, 7*(1), 63–92.

Orlikowski, W. J., & Hofman, D. J. (1997). An improvisational model for change management: The case of groupware technologies. *Sloan Management Review, 38*, 11–21.

Pentland, B. T., Hærem, T., & Hillison, D. (2011). The (n)ever-changing world: Stability and change in organizational routines. *Organization Science, 22*(6), 1369–1383.

Pettigrew, A. M., Woodman, R. W., & Cameron, K. S. (2001). Studying organizational change and development: Challenges for future research. *Academy of Management Journal, 44*(4), 697–713.

Pfeffer, J., & Salancik, G. R. (1978). *The external control of organisation.* New York: Harper and Row.

Phua, F. T. T. (2006). When is construction partnering likely to happen? An empirical examination of the role of institutional norms. *Construction Management and Economics, 24*(6), 615–624.

Rao, H., Monin, P., & Durand, R. (2003). Institutional change in Toque Ville: Nouvelle cuisine as an identity movement in French Gastronomy. *American Journal of Sociology, 108*(4), 795–843.

Robinson, W., Chan, P. W., & Lau, T. (2016a). Sensors and sensibility: Examining the role of technological features in servitizing construction towards greater sustainability. *Construction Management and Economics, 34*(1), 4–20.

Robinson, W., Chan, P., & Lau, T. (2016b). Finding new ways of creating value: A case study of servitization in construction. *Research-Technology Management, 59*(3), 37–49.

Schweber, L., & Harty, C. (2010). Actors and objects: A socio-technical networks approach to technology uptake in the construction sector. *Construction Management and Economics, 28*(6), 657–674.

Scott, W. R. (2001). *Institutions and organizations*. Thousand Oaks, CA: Sage.

Seo, M., & Creed, D. W. E. (2002). Institutional contradictions, praxis, and institutional change: A dialectical perspective. *Academy of Management Review, 27*(2), 222–247.

Sminia, H. (2011). Institutional continuity and the Dutch construction industry fiddle. *Organization Studies, 32*(11), 1559–1585.

Suchman, L. (1987). *Plans and situated actions*. Cambridge: Cambridge University Press.

Surachaikulwattana, P., & Phillips, N. (2017). Institutions as process. In A. Langley & H. Tsoukas (Eds.), *The SAGE handbook of process organization studies* (pp. 372–386). London: Sage.

Szentes, H. (2016). *Organizational tensions when managing interorganizational projects* (Unpublished doctoral thesis). Luleå University of Technology.

Szentes, H., & Eriksson, P. (2016). Paradoxical organizational tensions between control and flexibility when managing large infrastructure projects. *Journal of Construction Engineering and Management, 142*(4).

Thornton, P. H., & Ocasio, W. (2008). Institutional logics. In R. Greenwood, C. Oliver, K. Sahlin, & R. Suddaby (Eds.), *The SAGE handbook of organizational institutionalism* (pp. 99–129). London: Sage.

Thornton, P. H., Jones, C., & Kury, K. (2005). Institutional logics and institutional change in organizations: Transformation in accounting, architecture, and publishing. In C. Jones & P. Thornton (Eds.), *Transformation in cultural industries: Research in the sociology of organizations* (Vol. 23, pp. 125–170). Emerald Group Publishing Limited.

Tichy, N. M. (1983). *Managing strategic change*. New York: Wiley.

Tsoukas, H., & Chia, R. (2002). On organizational becoming: Rethinking organizational change. *Organization Science, 13*(5), 567–582.

Vaara, E., & Monin, P. (2010). A recursive perspective on discursive legitimation and organizational action in mergers and acquisitions. *Organization Science, 21*(1), 3–22.

Van de Ven, A. H., & Poole, M. S. (2005). Alternative approaches for studying organizational change. *Organization Studies, 26*(9), 1377–1404.

Weick, K. (1995). *Sensemaking in organizations.* Thousand Oaks, CA: Sage.

Weick, W. E., & Quinn, R. E. (1999). Organizational change and development. *Annual Review of Psychology, 50,* 361–386.

Whittington, R. (2017). Strategy as practice, process and institution: Turning towards activity. In A. Langley & H. Tsoukas (Eds.), *The SAGE handbook of process organization studies* (pp. 387–401). London: Sage.

Zairi, M., Letza, S., & Oakland, J. (1994). Does TQM impact on bottom line results? *TQM Magazine, 6*(1), 38–43.

6

Building Home Futures: Materialities of Construction and Meanings of Home in Self-help Building Practices

Monika Grubbauer

Introduction

Recent work in the social sciences concerned with the study of architecture has produced fascinating accounts of how buildings are lived in, used, and appropriated. Scholars have emphasized everyday practices, bodily movement, emotions, and affect in the interaction of people with buildings (Jacobs and Merriman 2011; Kraftl 2010). Similarly, within the field architectural history and theory, we can also note an increased interest into the ways in which practices of everyday use change architectural objects throughout time (Cupers 2013; Maudlin and Vellinga 2014). Others have continued to explore the ways in which buildings serve as symbolic agents and attest to the power of social actors in various ways, traditionally the power of the state and religious institutions, more recently iconic architecture which testifies to the power of transnational elites and private capital (Sklair 2006; Kaika 2010).

M. Grubbauer (✉)
HafenCity University, Hamburg, Germany
e-mail: monika.grubbauer@hcu-hamburg.de

© The Author(s) 2018
D. J. Sage and C. Vitry (eds.), *Societies Under Construction*,
https://doi.org/10.1007/978-3-319-73996-0_6

185

However, both strands of literature have largely limited their analyses to buildings *after completion* or on future visions of buildings *as to be completed*. Studies of ongoing "building work" in practices of dwelling and maintenance, for instance, take these practices to start only in the moment when buildings are inhabited and used (Strebel 2011; Jacobs and Cairns 2011). And social science studies of the symbolic dimension of buildings assume quite obviously that questions of representation and meaning only come into being with the completed buildings or circulated images thereof (Kaika 2011; Charney 2007; Broudehoux 2010).

In this contribution, I wish to focus on buildings which are by definition *under permanent construction*, the homes of self-help builders in the context of Latin American cities. These homes of low-income households are produced in self-organized building processes which easily span a couple of decades. In the industrialized construction sector, the building is a product produced for a market; it has to be finished at a certain point to be sold or rented out. In contrast, the homes of low-income households in the informal settlements of cities in Latin America, Africa, and Asia are built in processes of incremental financing and construction. These homes are mostly built for a lifetime, growing in dependence of the opportunities and needs of the households and meant to be passed on to children and grandchildren. They are sites of ongoing, multiple, and simultaneous practices of dwelling, construction, and livelihood generation.

Construction in self-help building might seem a special case without much relevance to the examinations of architecture and construction in European and North American contexts. In fact, the literature which deals with self-help building is largely situated in the field of development planning. There is a long tradition of research and practice concerned with self-help building practices which goes back to the 1960s and 1970s (Turner 1968, 1976). More recently, within the wider debates about urban informality in countries of the Global South, scholars have sought to deconstruct apocalyptic and dystopian narratives of the megacity (Gilbert 2007; Simone 2011) and to account for slums as "terrain[s] of habitation, livelihood and politics" (Roy 2011, p. 224). Still, the implicit assumptions remain that self-constructed homes are fundamentally different from residential architecture in Western contexts because of the incremental building processes that produce these buildings.

In this paper, I wish to argue against the "othering" of the homes of self-help builders in informal settlements in the Global South, more specifically in Latin America. I suggest that self-help construction provides valuable insights into the interplay between material affordances and representational powers of buildings which are of relevance for wider theoretical debates, not limiting it to the context of developing countries. Drawing on fieldwork conducted in Mexico City, I show how the materialities of construction and meanings of home in self-help building practices are interrelated. My key argument is that the ongoing construction of homes of self-help builders in Mexico is shaped by cultural norms and market structures principally in the same ways as in the context of residential architecture in the Global North. My discussion is based on fieldwork in Ecatepec, one of the oldest, largest, and most densely populated informal settlements in the metropolitan region of Mexico City. I visited and accompanied the architects of a non-profit enterprise working with households who seek to remodel and renovate their self-built homes. The insights presented are derived from in-depth interviews with members of this enterprise, informal conversations with home owners as well as observations made during site visits. The case study was part of a larger research project on linkages between the formal construction sector and informal building practices realized during a one-year research stay in Mexico City in 2014/2015.

The structure of the paper is as follows: In the first section, I will reflect on how self-help construction practices have been approached in social science debates and what the latest contributions to this debate are. In the second section, I will give some background on housing policies in Mexico; in the third, I proceed to discuss in more detail how the materialities of construction and the meanings of home are interrelated in the building of home futures by low-income households in Ecatepec. In the conclusion, I will reflect on what to learn from this discussion for theoretical debates about construction and societies.

Self-help Construction as Subject of Social Science Investigations

Scholarly work on marginalization and informality goes back to the 1960s and 1970s, many of the important early contributions originating in the Latin American context (Lomnitz 1975; Perlman 1976). The beginnings of the more specific work on self-help building are usually associated with John Turner (1968, 1976). However, Mexico has a particularly long and rich tradition of local research on the linkages between housing policies, informal settlements, and the autoconstruction of housing with work of authors such as Connolly (1979, 1985, 2009), Schteingart (1991, 2015), Garza and Schteingart (1978), and Bazant (1978, 2003) evolving over several decades. Based on the argument that "incremental building fits the livelihood strategies and conditions of the poor" (Ferguson and Smets 2010, p. 288), scholars and practitioners have for the past fifty years called for greater respect and sensitivity toward the needs and rationalities that drive self-help building processes. At the same time, there has been critique directed at Turner and others who have later built on his work, pointing out the danger of accepting self-help building as housing solution and thereby dismissing the responsibility of the state in providing affordable housing solutions for the poor (Connolly 1979; Burgess 1977; Conway 1982).

In most developing countries, the 1990s saw a turn to market-based solutions and a focus on the improvement of housing finance systems as part of the structural adjustment policies imposed by the World Bank and the IMF (Datta and Jones 1999a). Governments encouraged the development of private mortgage markets coupled with a strong emphasis on home-ownership. In Latin America, most governments introduced programs that combine savings, capital subsidies, and credits (so-called ABC schemes) to help the poor acquiring homes of their own, modeled on the Chilean example introduced in 1977 (Gilbert 2004; Klaufus 2010). However, a persistent problem is that these programs have tended to favor the middle-classes and have largely failed to improve the housing situation of the poorest groups (Bredenoord and Verkoren 2010; Klink and Denaldi 2014; Gilbert 2014). The reasons for this include the reluctance of the banking system to lend to the poor and the ultimately limited availability of capital subsidies. This has provoked sharp criticism of the ownership-centered housing policies of the past two decades (Campbell 2013; Lemanski 2011; Rolnik 2013). Moreover, the regularization and upgrading of informal settlements through land titling programmes has particularly come under attack. Since the 1990s, international development institutions have under the influence of Peruvian economist Hernando de Soto's *The Mystery of Capital* (2000) taken the stance that legal titles constitute the only viable option for households in informal settlements to secure their status, engage in housing improvement, and ultimately to escape poverty. Yet, research has shown that the positive effects of titling programmes are far from clear and, in many cases, perceived tenure security rather than land titles constitute the main motif for improvement of self-built dwellings (for an overview see Gilbert 2012).

Apart from the issue of property titles, among development institutions and many practitioners there is a consensus today that self-help building needs to be supported by increasing the speed and efficiency of the building processes (Ferguson and Smets 2010; Bredenoord et al. 2010; UN-Habitat 2005). In Mexico, there have been and continue to be many bottom-up initiatives by architects and housing activists who have been engaged in technical support of incremental building processes. However, this is usually either unpaid work or financed through donor organizations because self-help builders can normally not pay for the services of professional architects. Most recently, state-led housing policies attempt to institutionalize micro-loans as element of assisted self-help housing schemes. This allows for the payment of fees to architects but also brings the danger of further commercializing the microfinance sector through excessive profit margins and high interest rates which potentially lead to multiple borrowing and client over-indebtedness (Grubbauer, forthcoming).

Within the field of architecture, design, and planning, we can also note a renewed interest of professionals and educators in self-help strategies and incremental tactics of households in informal settlements. Professionals seek to work with local communities through a variety of participatory and co-design formats and tools. The aim of many of these initiatives is to improve living conditions of slum dwellers by developing design and planning approaches that are more sustainable and more adapted to local contexts (Rosa and Weiland 2013; Dovey 2013) but also to learn from informal urbanism in order to rethink design practice in Global North contexts. In all of this, informal urbanism is understood as "crucible of innovation" (Mehrotra 2012, p. xiv) rather than a condition in need of improvement. Part of this fascination with the informal by architects surely springs from the ambiguous status of buildings produced by incremental construction as "hybrid entity" (Rice 2015) between process and material form. Besides the highly publicized work of architects connected to leading Western architectural and planning schools concerned with informal urbanism (Brillembourg et al. 2005; Brillembourg and Klumpner 2013), we also find a growing number of non-academic organizations and initiatives supporting and realizing community-based design projects in the cities of the Global South.

There are, of course, considerable differences between the informal settlements in Africa, Asia, and Latin America not only in terms of how they come into being, in the ways in which authorities deal with them and in the extent of their social marginalization. They also differ in terms of built structures, building techniques, and the sourcing of building materials. In Mexico and much of Latin America, cement and concrete blocks constitute the most important basic building material. Cement consumption is continuously growing, the main driver for this being the housing sector, particularly informal construction. Based on data from the *Camara Nacional de Cemento* for the period 1994–2006, Fry calculates that between 40 and 50% of annual cement consumption come from informal construction (Fry 2013). Still, construction is a neglected issue in the research on self-help housing. There is certainly a lot of work currently concerned with vernacular building traditions and their potential in terms of more sustainable construction, up to the point of fetishizing local building materials (Grubbauer 2017). Yet there is too little research done in order to understand precisely how and where people source their materials for construction, how the markets for basic building materials are changing, and what kind of professional expertise self-builders needs to contract for the expansion and remodeling of their homes. This paper seeks to advance the debate in this respect.

Informal Settlements and Shifting Housing Policy Priorities in Mexico

About half of the population of 22 million inhabitants of the Metropolitan Region of the Valley de Mexico lives in so-called *colonias populares*, estimated 60% of the national housing stock is the product of self-help building. Self-help builders don't have access to the formal housing and mortgage market because of the lack of financial resources and formal employment. After invading a building plot or buying plots which often result from semi-legal or illegal subdivisions of land, the settlers engage in processes of incremental construction and financing as a way to secure affordable housing for their family (Ribbeck 2002). The *colonias populares* in Mexico City have been growing since the 1960s and are now inhabited in the second or third generation. Residential mobility in these quarters is extremely low. Peter Ward and his colleagues found in a comparison of household and ownership structures in several quarters in Mexico City over a period of thirty years that more than 80% of homes had not changed use or owner (Ward 2012).

Recent national housing policies in Mexico are more supportive toward self-help housing than those of earlier decades. Programs of the National Housing Commission (CONAVI) in cooperation with the Federal Mortgage Company (SHF) combine capital subsidies, micro-credits, and technical assistance in order to facilitate the improvement and remodeling of self-constructed homes of the lowest income groups. These programs have been introduced in Mexico over the past decade and constitute a response to the new Housing Law enacted in 2006 which for the first time recognized incremental building as legitimate housing strategy. More recently, these schemes have further been expanded in reaction to the negative consequences of the building boom of the late 2000s and the evident failures of market-based social housing provision through the industrialized construction sector (Monkkonen 2011, 2012; Puec-Unam 2012; Soederberg 2015). Awareness is growing that there is need to expand programs which serve to re-densify the consolidated informal settlements and to regenerate the self-constructed housing stock, especially in the Mexico City Metropolitan Area (Bredenoord and Cabrera 2014; Ward et al. 2015; Ziccardi and Gonzalez 2015).

The recent programs for improvement and remodeling of self-built homes implemented in Mexico, one the one hand, fulfill longstanding demands of housing activists and practitioners to take the needs of the urban poor more seriously adapt finance and assistance schemes to the needs of self-help builders (Datta and Jones 1999b; Bredenoord et al. 2014). On the other hand, these programs have also to be seen as response to market pressure and concession to interest of the finance and construction sector. The market for basic building materials used in self-help construction is both highly profitable and fairly stable in terms of demand and, thus, allows buffering the business cycles in the formal construction sector. For the past fifteen to twenty years, the large producers and retailers of building materials in Mexico have aggressively expanded into the market for low-income customers in both urban and rural areas. To attract clients and win their loyalty, technical assistance is often provided together with small consumer credits. Coupled with the state-led attempts to institutionalize housing microfinance as key instrument of national housing policies, housing microfinance is increasingly of commercial interest for the financial sector. As a result, the boundaries between commercial housing microfinance with excessive interest rates and the more socially oriented forms of housing microfinance are becoming increasingly blurred (Grubbauer, forthcoming).

Despite these deep ambivalences, it is justified to say that in Mexico, and in Latin America in general, there is currently a greater sensitivity for housing strategies of the poor than in earlier times. In Mexico City today the greater part of the city is made up by consolidated settlements which do not face the threat of eviction any longer and which have over the past decades been able to secure basic infrastructure. Even though the households in these settlements in many cases still lack legal land titles, they are engaged in continuous remodeling and expansion of their homes. These remodeling projects reflect current needs and prospects but also projections of home futures—how the needs and prospects of the households will develop in the future. How materialities of construction and meanings of home are interrelated in the making of these buildings is explored in the next part.

Materialities of Construction and Meanings of Home

Building processes of households in *colonias populares* in Mexico City proceed in dependence of the household situation: The needs of the families are balanced against their financial resources. About 60% of the labor force in Mexico is employed in the informal sector. Mexico is one of the countries in Latin America with particularly high and persistent informal employment. Informal workers lack job security, social protection, health insurance, and a contract that spells out their working conditions, hours, and wages. Under these conditions, any kind of investment beyond day-to-day expenses is difficult and some types of investment are simply not possible without some kind of additional financing. Money is borrowed from family and friends, also from community savings groups and through informal networks. In recent years, micro-credits for housing improvement and remodeling have become more important (Ferguson and Smets 2010; Smets 2012). Such housing microfinance products are offered by non-profit organizations and enterprises such as NGOs, savings, and loan cooperatives as well as a variety of for-profit banks and non-bank financial institutions, partly supported through programs of the national housing commission (CONAVI). In addition, consumer credits are provided by the large retailers (Walmart, Koppel, Home Depot) and the building material suppliers such as CEMEX and HOLCIM.

Apart from financial resources, also the labor needed for construction has to be sourced. Labor is always to a certain extent provided by households themselves and their family and friends based on relations of solidarity and reciprocity. Progress depends on the skills of household's members and on the amount of time that household members can invest or the support that they can mobilize. However, time is a precious good in Mexico City. People who live in *colonias populares* in Ecatepec travel two to three hours one way for their precarious jobs as guards, cleaning staff, or domestic workers in the more affluent areas in the southern and western parts of the city.

Beyond these informal sources for finance and labor, the contracting of additional labor, particularly professional masons, and the acquisition of building materials all rely on formal markets and monetized exchanges. In consequence, urban households in the metropolitan region of Mexico City are dependent on building suppliers, especially the producers and suppliers of basic building materials. Construction of self-built homes is largely cement-based, mostly in the form of concrete blocks which are made from Portland cement, water, and lightweight aggregates. The concrete blocks as well as other materials such as cement, sand, reinforcing steel, and iron sheets for roof cladding are sourced from local material shops. The market for these basic building materials is highly profitable and important for companies such as CEMEX, the world's largest producer of cement which dominates the Mexican market with 52% of the shares. The rationale is that the low-income housing market with its steady demand for basic building materials is less affected by the business cycles in the construction sector and can act as a buffer when demand in formal construction is low. The social and ecological costs of the expansion of production sites for basic construction material in rural areas, however, are immense. In central Mexico, landscapes have been profoundly transformed as farmers have converted their fields to mines for the surface mining of concrete aggregates (sand, gravel, crushed stone), especially the mining of lightweight aggregates such as tepetzil[1] used for the production of lightweight concrete blocks (Fry 2011).

For households, maintaining a permanent construction site poses all kinds of problems, and parts of the homes are not weatherproof; water tanks, sanitary facilities, electricity are introduced and adapted step-by-step; problems of mold and damp are common. A key challenge for urban households with plots of limited size is also posed by the logistics of handling the building materials. Households don't have cars, and transport has to be organized which is costly. In addition, households

---

[1]Faby Tuff, colloquially called tepetzil, is a volcanic air fall deposit extracted from surface mines which serves as desirable construction aggregate and is mostly processed into lightweight concrete blocks (Fry 2011).

tend to buy too large quantities of building materials because of being unsure about quantities; this however poses problems of storage, theft, and deterioration. Often, building materials become defective before they can be put to use.

On a national scale, the expansion of the markets for industrial building materials which cater to the needs and demands of self-help builders has become pervasive. Concrete blocks are found not only in urban areas but in all parts of the country. These blocks serve as vernacular building material substituting local materials such as wood or brick. This represents as geographer Matthew Fry argues a shift in "the use and perceptions of this building material. In this latest stage, blocks are no longer perceived as modernizing, but recognized for their functionality" (Fry 2008, p. 50). CEMEX in particular has been highly successful in promoting cement consumption within low-income groups through a "necessity narrative" which serves to normalize cement and prevents the transition to lower-carbon alternatives (Fry 2013). This has been facilitated by so-called bottom-of-the-pyramid business models which target low-income customers by offering packages of microcredits, technical assistance, and building materials.

This brings us to the other side of the picture, the meanings of home that drive and shape self-help building practices. Households act and plan according to "imagined futures" in which "houses are both concrete embodiments and imaginary representations of people's relations to their conditions of existence" (Holston 1991, p. 456). These home futures build on projections for how the family and income situation is going to develop but on also aspirations of social upward mobility and middle-class values. Such aspirations and values are mediated through migratory networks, mass, and social media and the promotional activities of multinational retailers and material suppliers (Lopez 2015; Klaufus 2012a).

More substantial and decisive remodeling projects typically consist of the construction of additional rooms or floors; these are decided upon depending on the number of family members which are planned to live in the home and contribute to the household's income. However, long-term planning of the future family situation including the number of required rooms and their layout is tricky as the family situation of urban poor households' tends to change quickly in dependence of work opportunities. Households often provide accommodation for extended family members on a temporary basis. The number of household members and the resulting needs are prone to change rather quickly because of the unstable nature of informal labor. Thus, households run the risk of misjudging the certainty of a future household scenario and precluding certain futures by deciding on a particular investment. This is even more serious, given the fact that self-built homes are typically built for a lifetime and meant to be handed over to children and grandchildren.

A typical strategy adopted by self-help builders in dealing with such uncertainties is to maximize the usable area. Builders accept that their home will stand half-finished for a long time in order to secure its usefulness at some point in a distant future, even though this future might never arrive. One problem of the state-subsidized upgrading projects which the non-profit enterprise that I examined is running in Ecatepec is that housing authorities have troubles in accepting this kind of logic. The regulations set by CONAVI require remodeling projects which receive state subsidies to deliver homes that are inhabitable and weather-resistant, i.e., have roof and windows. Thus, the kinds of home future imagined by self-help builders are disciplined through regulations which limit the floor space that is added and which require self-builders to accept mandatory expenses for doors, windows, and roofing. At the same time, the remodeling projects clearly have impact on the households living quality beyond what Green et al. describe as typical "project futures" (2012, p. 1643) of development aid and clients' satisfaction is reported to be extremely high.

Beyond questions of future household structure and income, home futures are also based on cultural values and aspirations of social upward mobility. Role models of middle- and upper-class suburban housing, especially US-Style, have enormous influence and influence layout, design and furnishing of Mexican homes. One reason for this is found in the extensive migrant networks which link Mexico with the USA. Architectural and urban historian Sarah Lynn Lopez (2015) shows in her book *The Remittance Landscape* that migrant remittances, dollars earned in the USA and sent to families and communities in Mexico, impact decisively on the built environment in the hometowns of the migrants. Mexico receives twenty-two billion US-Dollars per year in remittances. The remittance house, a home that is built or gets renovations and embellishments paid by migrant money, is an embodiment of the lifestyle, status, desires, and ambitions for migrant workers and their families, even when realized in their absence. Cultural anthropologist Christien Klaufus has also done fascinating work in showing how social status and architecture as well as physical and social mobility are linked in informal settlements working on cases from Ecuador (Klaufus 2012a, b).

The other reason for the importance of US-Style and international middle- and upper-class suburban housing is found in the presence of multinational retailers in Mexico. Walmart, by now the largest employer in the USA, entered the Mexican market in 1991 when it opened the first store outside the USA in Mexico City and has since then expanded successfully. It is the market leader in the retail sector with a network of 2400 stores. Home Depot which is the world's largest home improvement retailer currently operates more than sixty stores in Mexico and has become one of the largest retailers in Mexico since it entered the market in 2001. Home Depot increased its presence in Mexico in 2004, with the acquisition of Home Mart, the second largest Mexican home improvement retailer.

All of these multinational retailers are also very specifically targeting low-income customers, most importantly through consumer credits (Soederberg 2012). Producers of building materials such as CEMEX have also adopted micro-lending instruments as strategy to attract low-income customers, combining them with technical assistance. Moreover, to target the market of self-builders, CEMEX has invested since 2001 in establishing its own chain of material shops called Construrama. It operates through a franchising system with local distributors signing up for Construrama in exchange against training and equipment. The network comprises currently 3000 stores which specialize in everything needed for self-help construction. In many cases, the Construrama stores have been replacing local, owner-operated shops. CEMEX's Patrimonio Hoy program launched in 2000 is the most significant example for how the construction sector in Mexico has incorporated microfinance and its principle of social collateral as business strategy. Clients are organized in small savings groups, committed to weekly saving payments. In exchange for their savings, the clients receive building materials sourced from local Construrama shops and technical advice.

Despite such powerful processes of marketization, the making of the self-built homes in the urban peripheries of Latin America is more than the copy-paste of middle-class role models, this has been argued most forcefully by James Holston. In his seminal work on autoconstruction in Brazil, Holston argues that autoconstruction "generates an expansion of the field of the political" (Holston 1991, p. 447) by allowing the working classes to participate in modern consumer society while also engendering new political subjectivities. The aesthetic judgements which are visibly realized in self-built homes are crucial to allow "self images of competence and knowledge counter and replace those of disrespect and worthlessness" (ibid., p. 448). This in turn generates a sense of personal agency which is instrumental to develop political agency in residential mobilizations which build on collective action and participatory experience around issues of land, infrastructure, and home.

Conclusions

In Mexico today, we can observe powerful processes of internationalization, marketization, and financialization in the construction sector. These have affected self-help building practices over the past two decades in Mexico, replacing localized supply chains, increasing monetized forms of exchange, and changing cultural norms and values. These processes are facilitated by state-led national housing policies introduced a response to the new Housing Law enacted in 2006 which for the first time recognized incremental building as legitimate housing strategy. The respective programs of the National Housing Commission (CONAVI) and the Federal Mortgage Company (SHF) introduced in Mexico over the past decade combine micro-credits with technical assistance and capital subsidies in order to facilitate small-scale rehabilitation of informal settlements.

These state-led attempts to institutionalize housing microfinance as key instrument of national housing policies which aim at the rehabilitation of the self-help housing stock have to be seen in the light of the wider changes in the construction and retail sector in Mexico. The large producers and retailers of building materials in Mexico have been targeting the markets for low-income customers for the past fifteen years, greatly expanding their offer for consumer credits for housing improvement and remodeling. The rationale behind the expansion into these markets is that the two-thirds of the total housing stock in Mexico which are the product of self-help building constitute a vast and so far largely "underserved" market. Such estimates, however, not only reflect real demand but also normative assumptions about what type of built structure is regarded as completed. State-supported upgrading programs and credit schemes for home remodeling and self-help construction, no doubt, do serve financial interests keen on entering and expanding into low-income markets which in turn serves construction interests as demand for building materials is increased.

In this paper, I have sought to show that the "othering" of self-help building practices on the grounds of their incremental construction is misleading and even bears the danger of depoliticizing debates about urban informality and self-help building because it precludes an understanding two issues which are vital in the context of Mexico: First, how self-help construction of homes in informal settlements depends on the availability and affordability of industrial building materials and is shaped by ongoing processes of marketization and internationalization in the construction sector; second, how self-built homes are culturally hybrid and subject to a variety of influences but at the same time grounds for the development of new political and cultural subjectivities of the urban poor. Both, materialities of construction and meanings of home in informal building practices, are currently transformed by powerful processes of marketization as national and international building suppliers and retailers are increasingly targeting the market of self-help housing.

References

Bazant, J. (1978). *Tipología de vivienda urbana - analisis físico de contextos urbano habitacionales de la población de bajos ingresos en la ciudad de México*. México, D.F.: Editorial Diana.

Bazant, J. (2003). *Viviendas progresivas: construcción de vivienda por familias de bajos ingresos*. México, D.F.: Trillas.

Bredenoord, J., & Cabrera, L. (2014). Affordable housing for low-income groups in Mexico and urban housing challenges of today. In J. Bredenoord, P. Van Lindert, & P. Smets (Eds.), *Affordable housing in the urban global south: Seeking sustainable solutions*. London and New York: Routledge.

Bredenoord, J., Van Lindert, P., & Smets, P. (2010). Equal access to shelter: Coping with the urban crisis by supporting self-help housing. *Habitat International, 34*, 274–277.

Bredenoord, J., Van Lindert, P., & Smets, P. (Eds.). (2014). *Affordable housing in the global south. Seeking sustainable solutions*. London and New York: Routledge.

Bredenoord, J., & Verkoren, O. (2010). Between self-help—And institutional housing: A bird's eye view of Mexico's housing production for low and (lower) middle-income groups. *Habitat International, 34*, 359–365.

Brillembourg, A., Feireiss, K., & Klumpner, H. (2005). *Informal city: Caracas case*, München: Prestel.

Brillembourg, A., & Klumpner, H. (2013). *Torre David. Informal vertical communities*. Zürich: Lars Müller Publishers.

Broudehoux, A.-M. (2010). Images of power: Architectures of the integrated spectacle at the Beijing Olympics. *Journal of Architectural Education, 63*, 52–62.

Burgess, R. (1977). Self-help housing: A new imperialist strategy? A critique of the Turner school. *Antipode, 9*, 50–59.

Campbell, P. (2013). Collateral damage? Transforming subprime slum dwellers into homeowners. *Housing Studies, 28*, 453–472.

Charney, I. (2007). The politics of design: Architecture, tall buildings and the skyline of central London. *Area, 39*, 195–205.

Connolly, P. (1979). Autoconstrucción espontánea: solución o problema? *Vivienda, 4*, 144–153.

Connolly, P. (1985). The politics of the informal sector: A critique. In N. Redclift & E. Mingione (Eds.), *Beyond employment, household, gender and subsistence*. Oxford: Blackwell.

Connolly, P. (2009). Observing the evolution of irregular settlements: Mexico City's colonias populares, 1990 to 2005. *International Development Planning Review, 31,* 1–35.

Conway, D. (1982). Self-help housing, the commodity nature of housing and amelioration of the housing deficit: Continuing the Turner-Burgess debate. *Antipode, 14,* 40–46.

Cupers, K. (Ed.). (2013). *Use matters: An alternative history of architecture.* London and New York: Routledge.

Datta, K., & Jones, G. A. (1999a). From self-help to self-finance: The changing focus of urban research and policy. In K. Datta & G. A. Jones (Eds.), *Housing and finance in developing countries.* London and New York: Routledge.

Datta, K., & Jones, G. A. (Eds.). (1999b). *Housing and finance in developing countries.* London and New York: Routledge.

Dovey, K. (2013). Informalising architecture: The challenge of informal settlements. *Architectural Design, 83,* 82–89.

Ferguson, B., & Smets, P. (2010). Finance for incremental housing; current status and prospects for expansion. *Habitat International, 34,* 288–298.

Fry, M. (2008). Mexico's concrete block landscape: A modern legacy in the vernacular. *Journal of Latin American Geography, 7,* 35–58.

Fry, M. (2011). From crops to concrete: Urbanization, deagriculturalization, and construction material mining in central Mexico. *Annals of the Association of American Geographers, 101,* 1285–1306.

Fry, M. (2013). Cement, carbon dioxide, and the 'necessity' narrative: A case study of Mexico. *Geoforum, 49,* 127–138.

Garza, G., & Schteingart, M. (1978). *La acción habitacional del Estado de México.* México, D.F.: El Colegio de México.

Gilbert, A. (2004). Learning from others: The spread of capital housing subsidies. *International Planning Studies, 9,* 197–216.

Gilbert, A. (2007). The return of the slum: Does language matter? *International Journal of Urban and Regional Research, 31,* 697–713.

Gilbert, A. (2012). De Soto's the mystery of capital: Reflections on the book's public impact. *International Development Planning Review, 34*(3), v–xviii.

Gilbert, A. G. (2014). Free housing for the poor: An effective way to address poverty? *Habitat International, 41,* 253–261.

Green, M., Kothari, U., Mercer, C., & Mitlin, D. (2012). Saving, spending, and future-making: Time, discipline, and money in development. *Environment and Planning A, 44,* 1641–1656.

Grubbauer, M. (2017). In search of authenticity: Architectures of social engagement, modes of public recognition and the fetish of the vernacular. *City, 21*(6): 789–799.

Grubbauer, M. (Forthcoming). Assisted self-help housing in Mexico: Advocacy, (micro)-finance and the making of markets. *International Journal of Urban and Regional Research*.

Holston, J. (1991). Autoconstruction in working-class Brazil. *Cultural Anthropology, 6*, 447–465.

Jacobs, J. M., & Cairns, S. (2011). Ecologies of dwelling: Maintaining high-rise housing in Singapore. In S. Watson & G. Bridge (Eds.), *The new companion to the city*. Oxford: Blackwell.

Jacobs, J. M., & Merriman, P. (2011). Practising architectures. *Social and Cultural Geography, 12*, 211–222.

Kaika, M. (2010). Architecture and crisis: Re-inventing the icon, re-imag(in)ing London and re-branding the City. *Transactions of the Institute of British Geographers, 35*, 453–474.

Kaika, M. (2011). Autistic architecture: The fall of the icon and the rise of the serial object of architecture. *Environment and Planning D: Society and Space, 29*, 968–992.

Klaufus, C. (2010). The two ABCs of aided self-help housing in Ecuador. *Habitat International, 34*, 351–358.

Klaufus, C. (2012a). Moving and improving: Poverty, globalisation and neighbourhood transformation in Cuenca, Ecuador. *International Development Planning Review, 34*, 147–166.

Klaufus, C. (2012b). The symbolic dimension of mobility: Architecture and social status in Ecuadorian informal settlements. *International Journal of Urban and Regional Research, 36*, 689–705.

Klink, J., & Denaldi, R. (2014). On financialization and state spatial fixes in Brazil. A geographical and historical interpretation of the housing program My House My Life. *Habitat International, 44*, 220–226.

Kraftl, P. (2010). Geographies of architecture: The multiple lives of buildings. *Geography Compass, 4*, 402–415.

Lemanski, C. (2011). Moving up the ladder or stuck on the bottom rung? Homeownership as a solution to poverty in urban South Africa. *International Journal of Urban and Regional Research, 35*, 57–77.

Lomnitz, L. A. (1975). *Cómo sobreviven los marginados*. S.A.: Siglo XXI de España Editores.

Lopez, S. L. (2015). *The remittance landscape. Spaces of migration in rural Mexico and urban USA*. Chicago and London: University of Chicago Press.

Maudlin, D., & Vellinga, M. (Eds.). (2014). *Consuming architecture: On the occupation, appropriation and interpretation of buildings*. London and New York: Routledge.

Mehrotra, R. (2012). Foreword. In F. Hernández, P. Kellett, & L. K. Allen (Eds.), *Rethinking the informal city. Critical perspectives from Latin America*. New York and Oxford: Berghahn.

Monkkonen, P. (2011). Do Mexican cities sprawl? Housing-finance reform and changing patterns of urban growth. *Urban Geography, 32*, 406–423.

Monkkonen, P. (2012). Housing finance reform and increasing socioeconomic segregation in Mexico. *International Journal of Urban and Regional Research, 36*, 757–772.

Perlman, J. E. (1976). *The myth of marginality: Urban poverty and politics in Rio de Janeiro*. Berkeley and Los Angeles: University of California Press.

Puec-Unam. (2012). *México. Perfil del sector de la vivienda*. México, D.F.: Universidad Nacional Autónoma de México/Programa Universitario de Estudios sobre la Ciudad.

Ribbeck, E. (2002). *Die informelle Moderne—Spontanes Bauen in Mexiko-Stadt*. Heidelberg: Awf.

Rice, L. (2015). Informal architecture/s. In R. Louis & D. Littlefield (Eds.), *Transgression: Towards an expanded field of architecture* (pp. 87–101). London and New York: Routledge.

Rolnik, R. (2013). Late neoliberalism: The financialization of homeownership and housing rights. *International Journal of Urban and Regional Research, 37*, 1058–1066.

Rosa, M. L., & Weiland, U. (Eds.). (2013). *Handmade urbanism: From community initiatives to participatory models: Mumbai, São Paulo, Istanbul, Mexico City, Cape Town*. Berlin: Jovis.

Roy, A. (2011). Slumdog cities: Rethinking subaltern urbanism. *International Journal of Urban and Regional Research, 35*, 223–238.

Schteingart, M. (1991). *Espacio y vivienda en la ciudad de México*. México, D.F.: Centro de Estudios Demográficos y de Desarrollo Urbano, Colegio de México.

Schteingart, M. (2015). *Desarollo urbano-ambiental, políticas sociales y vivienda: Treinta y cinco años de investigación*. México, D.F.: El Colegio de México.

Simone, A. (2011). The ineligible majority: Urbanizing the postcolony in Africa and Southeast Asia. *Geoforum, 42*, 266–270.

Sklair, L. (2006). Iconic architecture and capitalist globalization. *City, 10*, 21–47.

Smets, P. (2012). Housing policies in developing countries: Microfinance. In S. J. Smith (Ed.), *International encyclopedia of housing and home*. San Diego: Elsevier.

Soederberg, S. (2012). The Mexican debtfare state: Dispossession, micro-lending, and the surplus population. *Globalizations, 9,* 561–575.

Soederberg, S. (2015). Subprime housing goes south: Constructing securitized mortgages for the poor in Mexico. *Antipode, 47,* 481–499.

Strebel, I. (2011). The living building: Towards a geography of maintenance work. *Social and Cultural Geography, 12,* 243–262.

Turner, J. C. (1968). Housing priorities, settlement patterns, and urban development in modernizing countries. *Journal of the American Institute of Planners, 34,* 354–363.

Turner, J. F. C. (1976). *Housing by people: Towards autonomy in building environments*. London: Marion Boyars.

Un-Habitat. (2005). *Financing urban shelter. Global report on human settlements*. London: Earthscan.

Ward, P. M. (2012). "A patrimony for the children": Low-income homeownership and housing (im)mobility in Latin American cities. *Annals of the Association of American Geographers, 102,* 1489–1510.

Ward, P. M., Jiménez Huerta, E. R., & Di Virgilio, M. (Eds.). (2015). *Housing policy in Latin American cities. A new generation of strategies and approaches for 2016 UN-Habitat III*. London and New York: Routledge.

Ziccardi, A., & Gonzalez, A. (Eds.). (2015). *Habitabilidad y política de vivienda en México*. México, D.F.: Universidad Nacional Autónoma de México.

7

From Relational to Regressive Place-Making: Developing an ANT Theory of Place with Housebuilding

Daniel J. Sage and Chloé Vitry

The term 'actor-network' was first coined by Michel Callon in 1986 to explain how social studies of technoscience might take better account of the socially transformative role of technologies and science. Callon's (1986a) purpose was to address an analytical challenge: How can social scientists understand technological innovations, when the concepts they typically deploy (e.g. political interest, class and social capital) fail to acknowledge that technologies are always 'rebuilding society by introducing unpredictable variations and new associations' (Callon 1986a, p. 21). That is to say, if only social contexts and concepts populated by humans are deployed to help us understand technoscience, we simultaneously reduce the complexity, heterogeneity and dynamics of both technoscience and society (cf. Latour 1987, 2005). Instead, Callon (1986a)

D. J. Sage (✉)
School of Business and Economics, Loughborough University, Loughborough, UK
e-mail: d.j.sage@lboro.ac.uk

C. Vitry
School of Business, University of Leicester, Leicester, UK
e-mail: cav5@le.ac.uk

© The Author(s) 2018
D. J. Sage and C. Vitry (eds.), *Societies Under Construction*,
https://doi.org/10.1007/978-3-319-73996-0_7

205

proposed an alternative form of analyses where technoscience and society are treated more symmetrically (see also Callon 1986b). Central to Callon's (1986a) approach is the notion that a technoscientific network builder must construct *simplified* images of heterogeneous actors, whether people, organizations, technologies or cities, to enrol them together in support of technology (Callon 1986a). Stemming from its original emphasis on symmetry, network stabilization and actor simplification, for over thirty years actor-network theory (ANT) has been critiqued for reducing difference, for ignoring geographical (Woods 1998) and historical (Mutch 2002) particularities, for collapsing differences between people and objects (Collins and Yearley 1992; Whittle and Spicer 2008), for obscuring differences between social groups (Ettlinger 2003; Lee and Brown 1994; Star 1991), and even eviscerating novelty from social analysis (Thrift 2000; Müller and Schurr 2016), in short, for being totalitarian (Latour 1999).

Lending tacit support to these criticisms, ANT has often, though certainly not always (Gadd 2016; Hitchings 2003; Lees and Baxter 2011) been empirically deployed to understand rather homogenizing geographies, whether the emergence of world city networks (Smith 2003), the relations between transnational firms (Jones 2008) or the global power projections of corporations, governments and NGOs (Allen 2011a; Müller 2014). In contrast, ANT has more fleetingly been empirically related to geographies more commonly associated with intensive variations and differences, such as bodies (Müller 2015), landscapes (Allen 2011b), places (Gadd 2016; Hitchings 2003) and emotions/affects (Lees and Baxter 2011; Müller and Schurr 2016). Concomitant with this state of affairs, across two decades of varied geographical engagements with ANT, it is difficult to discern any sustained and explicit ANT theory of place as opposed to the various studies that discuss ANT and space (Amin 2002; Law 2002a; Murdoch 1998; Müller 2015). This is somewhat surprising because from its birth, actor-network theory has always been sensitive to intensive spatial heterogeneities, that is to place.

Our purpose in this chapter is to rehabilitate this latent treatment of place within ANT, in order to develop and consider ANT's potential contribution to theories of place, especially regarding the relationality and eventfulness of place-making (Castells 2000; Massey 1991, 2005;

Pierce et al. 2011; Sack 1997). In keeping with the scope of this edited collection, our argument is theoretically constructed and then refined empirically alongside an analysis of the early stages of the Garendon Park housebuilding construction project located in Loughborough in the English East Midlands. To further introduce our argument for an ANT theory of place, we will briefly return, in the first section of this chapter, to Callon's (1986a) founding use of the term actor-network—the development of a specific electric vehicle in France in the 1970s by EDF, in conjunction with an energy company, CGE, the French ministry of transport, various municipalities and Renault. Through close examination of this seminal ANT study, we can start to consider how place might already be apprehended in ANT scholarship.

The Place of Place in Actor-Network Theory

In his foundational ANT study, Callon (1986a) identifies EDF as the network builder, and their electric vehicle as the transformative solution to the problems of, amongst various actors, French city and town municipalities. For EDF, the interests of the municipalities in this project stem solely from the value of this electrical vehicle to be used as public transport, reducing urban noise and pollution: '[for] EDF, the city is reduced to the city-council-that-wants-to-preserve-the-town-centre-at-all-costs' (Callon 1986a, p. 30). This place simplification where the municipality is to be changed by technoscience, in turn, requires that the municipal government stabilizes all manner of relationships it has with other actors: 'the middle-class electorate that trusts it, the pedestrian precinct that pushes the flow of traffic to the edge of the town centre, the urban spread, and the system of public transport which enables the inhabitants of the suburbs to come and do their shopping in the town centre' (p. 30). Callon (1986a) is mindful of the limits of such place simplifications: 'towns consist of more than public transport, the wish to preserve town centres and town councils that constitute their spokespeople. They differ from one another with respect to population, history and geographical location. They conceal a hidden life whose anonymous destinies interact' (p. 29). If these complex actors come to the fore and reject the simplification of the municipality prescribed by EDF, then their electric vehicle cannot transform society; if, for example, it appears 'the town council is not representative; living conditions in different neighbourhoods cannot be reduced to those in the town centre; and the system of public transport is but one aspect of a large urban structure' (Callon 1986a, p. 29). Hence, the fate of EDF's electric vehicle, or any actor looking to change the world, is said to be tied to practices wherein actors accept or not simplified images of their identities and interests provided by network builders, and then simultaneously accept their role in a wider network. This process is, in turn, tied to their own capacity to simplify and juxtapose constitutive elements they seek to represent in their own networks and so on.

Callon (1986a) offers the metaphorical concept of the 'black box[1]' to explain this process wherein 'each element is part of the [embedded] chain that guarantees the proper functioning of the object. It can be compared to a black box which contains a network of black boxes that depend upon one another both for their proper functioning and the functioning of the whole' (p. 31). It is here, in these infinitely regressive movements between two or more simplifying and stabilizing, or 'black boxed', actor-networks, that eventfulness, or agency, seems to reside for Callon (1986a). This is because actors' capacities to mould and shape others (Callon 1986a, p. 32), via simplifications, depends, in turn, on the capacities of these other actors to produce images that simplify yet more actors. Agency remains possible because, all of these images, all of these nested and regressively simplifying movements, can, in turn, not only influence, but also transform, a larger, more connected, actor-network. This is because each actor-network is only as strong as the capacity of any actor to enrol its most resistant link (Callon 1986a, p. 23)

---

[1]The term 'black box', while used by Callon (1986a), is developed more in Latour (1987, p. 2). A 'black box' describes the outcome of a stabilized actor-network; that is an actor, such as technology or scientific fact, whose relational complexity has been stabilized so that the actor exists in terms of a predictable set of inputs and outputs. The process of constructing a closed 'black box' thus corresponds with the end process of 'punctualization' in network building described by Callon (1991) or 'mobilization' in Callon (1986b). Importantly, for Latour (1987, pp. 78–80), 'black boxes' can also be opened up and disputed, although this process is resource intensive and requires the recruitment of more 'black boxes'.

and these resistances must be tested in practice (Callon 1986a, p. 32). Moreover, as Callon (1986a) explains, it is through these infinitely regressive processes of simplification and juxtaposition—where, what he would later term 'punctualized' (Callon 1991, p. 53) networks are nested inside each other—that we come to understand how societies are transformed (as in Callon 1986a, b).

However, while the simplification and juxtaposition of actors, the reduction of complexity, are the aim of EDF, and perhaps all technoscientific network builders (Callon 1986a), this group of actors must also be highly sensitive to difference and heterogeneity. It is at this juncture that we can start to flesh out how we can approach place: ANT suggests that the relational complexities of place, indeed of all difference, can only be simplified through a measure of sensitivity to what is excessive to that simplification process. That is to say, it is only through receptivity to difference that network builders can produce different images of actors that are then simplified, and thus easily transported and combinable, yet still capture enough of the different agendas, interests and capacities of the actors involved. For EDF to be successful it would seem, contra Callon (1986a, p. 29), that it would benefit them to know more, not less, about other actors, and especially their variegated capacities to simplify the interests of others within their own constitutive networks. For example, if the municipal government in Callon's (1986a) study wants to understand the varied interests of the middle class that may support it, it may benefit that government to understand the capacity of spokespeople for the middle class to simplify the interests of different groups within the middle class. After all, if any one of these simplifications fails, then the middle class cannot be treated as a simplified whole, and the work of the municipal government, and, in turn, EDF, to enrol them to support their electric vehicle may also fail.

To be clear at this juncture, while this regressive network building process may be homogenizing this is only the case vertically not horizontally. This is because the simplified image of an actor in an actor-network is certainly not homogeneous with all the other actors in that network. Indeed, as Callon (1986a) explains, each actor fulfils an indispensable (i.e. different) function in the network (Callon 1986, p. 23). And so homogenization, where it appears, is vertical, corresponding to the congruence drawn, or not, between actors and the simplified images presented to them by network builders seeking to enrol them in their networks. This is precisely why even from its conception, the elision of difference, including placial difference, is not, contra numerous claims (Lee and Brown 1994; Whittle and Spicer 2008), a defining aspect of ANT thinking. Rather, ANT provides a distinctly *regressive* ontology of difference, which, as will be discussed in due course, is far from antithetical to concepts of place-making as relational and eventful (Cresswell 2015; Pierce et al. 2011). Callon (1986a) draws our attention to the notion that a place, like any actor, can only exist as such, that is as distinct from, yet related to, another, if it works with some aspect of simplification of itself by another actor. Without simplification, a place, perhaps a town or city, collapses into an infinite regress: the indeterminate complexities of its constituent actors and relations, and their multiple images and constructions (Callon 1986a, pp. 28–29). But crucially, this simplification process, wherein a town or city is knowingly reduced in its complexity, exemplified perhaps by the rise of place branding in planning and development discourse (Kavaratzis and Ashworth 2005), also provides an opportunity for actors to influence images of place simplification and thus possibilities for eventfulness, for agency.

In the remainder of this chapter, we seek to further elaborate on the conceptual contours and novelty of thinking place with ANT. In so doing, we draw particular attention to how the dynamic between simplification and infinite regressive complexity within ANT, as vividly drawn by Callon (1986a), offers a valuable starting point to develop a distinctive, and novel, approach to understanding place. To this end, we develop our approach through empirical analysis of a housebuilding project in the English East Midlands. UK housebuilding offers an especially germane domain to an ANT analysis of place-making. Housebuilding companies are now formally encouraged in national planning policy to 'establish a strong sense of place, using streetscapes and buildings to create attractive and comfortable places to live, work and visit' (NPPF 2012; see also Building for Life 2015, p. 5). However, well-established industry pressures towards profit-maximization through design standardization, especially on volume housebuilders (Nicol and Hooper 1999), impose significant constraints of the capacity of housing developers to create the sorts of distinctive, contextualized and diverse places encouraged in government-endorsed design and planning guidance (such as Building for Life 2015). In other words, across most housing companies, developments are constructed using a spatial language where the particularity of 'place' is central, but these places are often, if not always, constructed out of highly simplified, singular, stereotypical images of place—branded places (Kriese and Scholoz 2012).

Before turning to our empirical context, we first broaden and deepen our consideration of an ANT theory of place by comparing Callon's (1986a) largely overlooked, yet rather implicit, sensitivity to place with the wider development of ANT thinking towards place (and space), especially within human geography. We then locate a latent ANT approach to relational place alongside recent work in human geography on relational place-making (Cresswell 2015; Pierce et al. 2011). These concepts are then developed further through our empirical study in order to develop an ANT treatment of place and in particular notions of the eventfulness as well as relationality. We conclude by considering some of the contributions ANT might make to burgeoning debates on relational place-making.

Elaborating ANT Theories of Place

Despite the popularity of ANT in human geography (Müller 2015), key ANT exponents, such as Michel Callon, Bruno Latour and John Law, have only intermittently discussed the implications of their approach for concepts of 'space' and even less so 'place'. One of the most sustained considerations of spatialities and ANT can be found in related studies by Law (1986) and Latour (1987). Across these studies, these two seminal ANT thinkers develop their relational approach with reference to terms such as 'space' and 'place' through the study of the socially transformative technology of the Carrack sailing vessel as it emerged in fifteenth- and sixteenth-century Portugal. Paralleling the study of Callon (1986a), Law (1986), and Latour (1987) explain how a technology, the Carrack, modified (Portuguese) society by creating new associations between humans and non-humans, specifically: lateen and square rigs, large cannons, larger and stronger hulls, logbooks, work contracts, astronomical navigation instruments and documents, charts and trained crews. As these actors were simplified and juxtaposed by the network builder—the Portuguese government—Portuguese ocean-going vessels became highly durable and mobile, acting at a distance to control colonial trade and possessions (Law 1986; Latour 1987). Here, Callon's (1986a) early study of technoscientific place-making is complemented by a colonial variant: 'How to act at a distance on unfamiliar events, *places* and people? Answer: by somehow bringing home these events, places and people' (Latour 1987, p. 223; emphasis added). In these colonial analyses, placial particularities, the distant islands, reefs and coastlines of the Atlantic and Indian Oceans, are simplified and juxtaposed—enabling them to become *mobile, stable* and *combinable* (Latour 1987)—as they are translated into longitudinal and latitudinal coordinates on navigational maps. These places then arrive together in Lisbon—conceived as a 'centre of calculation' (Latour 1987)—to create an 'obligatory passage point' (Callon 1986b) for those at the 'margins', looking to trade, to navigate, to prosper, to be secured (Law 2002b, p. 152). As Law (2002b) explains: 'these are the obligatory passage points fashioned to be the center of the political universe, the *places* through which everything is made to pass' (p. 152, emphasis added). But the spatial effect of this centring does not simply homogenize places, rather:

> The effect is to produce a distribution between center and periphery, and to efface the possibility that there are other locations that might escape the gravitational pull of that center – or, indeed, the possibility that the world might perform itself without the need of special centers. (Law 2002b, p. 153)

By considering how place-based differences, such as 'center/periphery' (Latour 1987; Law 2002a, b), are relationally produced, these seminal ANT studies suggest that there exists no essential opposition between universal and local knowledge, between unbounded, geometric, space and bounded, particular, place; this is because all spatialities, even geometric space, global panoramas, are always bounded, produced somewhere, in some*place* (Latour 2005, p. 183), and bounded through certain relations (Latour 1987, p. 230). Even if we view a place, with Latour (1987), as a 'centre of calculation' that can reproduce distant places locally, plot them on gridlines and govern them at a distance, these are still bounded places, they are bounded by their networks: 'as long as the Portuguese carrack disappeared en route, no space beyond the Bojador Cape could be pictured' (Latour 1987, p. 230). Furthermore, for Latour (1987) and Law (1986), it is clear that placial network building may be far less consensually negotiated than Callon (1986a) suggests—simplified images of other actors, including places, are constructed that other actors must accept or risk annihilation. For instance, Law (1986, p. 6) describes how the Portuguese constructed an image of Eastern peoples around their capacity for conformity and submission to European supremacy—enforced through the defensive castles, small arms and large cannon on Portuguese Carracks—denying this placial ordering risked death (Law 1986, p. 239).

These early ANT treatments of space and place, and attendant concepts such as 'centre of calculation' and 'centre/periphery', have been developed across various engagements with network spatialities, in human geography and beyond. In a much-cited paper on the subject, Murdoch (1998) discusses how ANT figures space in terms of prescription and negotiation: 'even in spaces framed by formal modes of calculation there is some scope for negotiation, that is, actors can carve out for themselves a degree of autonomy from the network prescriptions' (p. 363). Hence even when simplified images are prescribed to enrol places in networks, for example, a town accepts an image of it requiring an electric vehicle to solve its problems (Callon 1986a), a great deal cannot be prescribed and will remain negotiable as not all of the town's constituent elements, its diverse middle-class population, can be, or perhaps needs to be, prescribed in such a simplified image. As Murdoch (1998) explains: 'the network can be acted out at precisely the same time as its precepts are flouted; actors thereby become conformists and non-conformists simultaneously' (p. 366). Significantly, here (placial) particularity, for example, a town or colonized nation that does not conform to the simplified image offered by a network builder, does not so much negate the entire network, heralding its collapse (as suggested in Callon 1986a, b; Latour 1987; Law 2002b), rather its uniqueness is partly produced by the network and its incapacity to prescribe. Murdoch (1998) thus supplements the placial particularities implicit within early ANT thinking. Specifically, he argues that such place-making does not only correspond with intentionally produced differences between heterogeneous, yet simplified, actors and forces (as suggested by Callon 1986a, b; Latour 1987; Law 1986), but also novel, emergent, differences precipitated by the limitations of the processes of simplification and stabilization that bind networks together. Two empirical studies are illustrative of these processes. Graham (1998) offers the case of ICT use: 'the use of faster and faster telematics systems actually increases the demands for face-to-face contact so that the interpretive loads surrounding information glut can be dealt with rapidly and competitively' (p. 180). In other words, here the spatial standardization of a centre of calculation, the flow of standardized data, not only produces greater placial variations (ad hoc face-to-face communication), but its flow is conditional on those variations. Müller (2014) similarly explains, through his study of Olympic organizing, how centres of calculation unavoidably produce placial variations and differences:

> … despite the elaborate apparatus for capturing and circulating knowledge that the IOC [International Olympic Committee] has developed, some knowledge continues to escape being brought back home, but rather creates separate flows of knowledge, flows bypassing the IOC. These bypasses limit the possibilities of enrolment since alternative sources of knowledge exist to fill the need for knowledge and shape action and, as a consequence, enrolment often remains partial and selective. (p. 336)

In the light of our review of ANT scholarship, we can start to recognize how ANT contains two particular, albeit related, versions of how relations can foster placial variation. Firstly, places can exist in their particularity as bounded, yet relationally constituted, 'centres' and 'peripheries' in geometric spaces (Latour 1987; Law 2002a, b). Here, as with Callon (1986a), the uniqueness of place is constituted by the intentional establishment, even imposition, of simplified images of place. These entities are defined by nested and regressive actor-networks that relationally delineate placial boundaries and differences, including in geometric spaces (e.g. Allen 2011a; Gadd 2016; Law 2002a; Jones 2008; Smith 2003; Young 2010). Secondly, placial variations can be created in the eventful juxtapositions that occur when network builders attempt to simplify and bound the particularity of place in geometric space and in the process collide and entangle with other relations, including other actor-networks (Callon 1986a; Graham 1998), more fluid spatialities (see e.g. Amin 2002; Bear and Eden 2008; Hitchings 2003; Müller 2014) and affective states (Gadd 2016; Lees and Baxter 2011; Müller and Schurr 2016). These juxtapositions may imbue emergent transformations, or impose limits, on the place simplifications of network builders, even rendering those strategies as conditional on the eventfulness of place (see Bear and Eden 2008; Müller 2014).

Locating an ANT Approach to Relational Place-Making

Thinking placial differences with ANT helps challenge romantic views of the particularity of meaningful, bounded places opposed to the standardizing, abstract, global connectedness of space—for ANT, all differences, placial or otherwise, are always generated relationally (Callon 1986a; Latour 2005; Law 2002a). However, the rejection of romantic, fixed, treatments of place is hardly unique to ANT; indeed relational approaches to place are long established in human geography and beyond (Casey 1998; Castells 2000; Malpas 1999; Massey 1991, 2005; Pred 1984; Thrift 1996; Sack 1997). Thus, before introducing our empirical study, we will discuss the extent to which ANT's latent treatment of place offers a *novel* approach to thinking place relationally. Rather than comprehensively review this extensive literature on relational place-making (see Antonisch 2011; Cresswell 2015; Pierce et al. 2011), we focus our attention on the novelty of two specific strands of thinking about place derived from ANT as mapped out so far in this chapter: (i) places as simplifications produced by network builders to arrest potentials for regressions; and (ii) places as outcomes of the excess of eventfulness accompanying those simplifications.

To be sure, ANT does mirror various recent relational approaches to place wherein we are to understand 'place as a combination of parts that links the inside of a place to what lies beyond – places as syncretic sites of gathering and assemblage' (Cresswell 2015, p. 55). Concepts of places as a *gathering* and *assemblage* clearly dovetail with ANT in considering how places are composed as distinct relations between human and non-human entities:

> [Consider] the place you call home, you can see that there are forces at work that stabilize its identity, ranging from the legal structures that make it your house to the physical forces that hold the walls, floors, and ceilings together. You can also think of processes that might make it less stable – the natural forces of entropy or the lines that lead out from the home to the wider world. (Cresswell 2015, p. 54)

Cresswell's (2015) treatment of place parallels the emphasis given in ANT to how the relations between heterogeneous actors align, or not, to bring into being new ontological entities, whether houses or towns. We contend the specific novelty of ANT corresponds not with its insistence on the relationality of place-making per se, but with two further elements: *network building* and *simplification*. Regarding the first point, our reading of place with ANT differs somewhat from some relational notions of place as a *gathering* and *assemblage* by describing in more detail how a calculative actor—a network builder—can seek, even covertly or coercively, to assemble a place as an obligatory passage point (Callon 1986b) or 'centre for calculation' for other actors (Latour 1987; Law 1986). Cresswell (2015) explains how the concept of an assemblage is more limited in this regard: 'The ways in which parts are combined in an assemblage are not structurally necessary or preordained. They are not directed by some higher force. Their combination is contingent' (p. 53). By contrast, for ANT, relations can be (though certainly are not always—Law 2002a), structurally necessary: 'it [the actor-network] still remains only as strong as its weakest link no matter how grandiose some of its elements may be' (Latour 1987, p. 124).

To unpack the unique contribution of ANT further, we can compare it to Pierce et al.'s (2011) framework of relational place-making (see also Pierce and Martin 2015). Pierce et al. (2011) describe a networked concept of place-framing which is reminiscent of ANT as it is concerned with how networked negotiations between human and non-human objects (e.g. buildings, social groups, roads, parks, forests and corporations), and their space-time trajectories, are drawn together by individual people to produce 'place bundles' that serve strategic political and social ends (Pierce et al. 2011; Pierce and Martin 2015). Pierce et al. (2011) even fleetingly describe their 'place bundles' as conglomerations of objects that are themselves 'actor-networks' (p. 58). Their use of the term 'framing' also mirrors Callon's earlier use of the same term to describe the processes of simplification (Callon 1999, p. 188). Finally, the notion of place-framing as an ongoing achievement, with no micro-/macro-scale starting point, where heterogeneous actors interact around multiple place frames, even to foster actor multiplicities (Pierce et al. 2011; Pierce and Martin 2015), all resonates well with ANT (Callon 1986a, 1999; Latour 1987). However, despite Pierce et al. (2011) introducing a useful analytical distinction between place-framing producers/reproducers, and blenders/decision makers, their analyses of the specific practices by which (human) network builders might order places by enrolling actors in support of their stabilizations are far less detailed than even early ANT analyses (e.g. Callon 1986; Latour 1987). Latour (1987, pp. 108–121), for example, lists a myriad of network building strategies: defining shared interests, isolation from extant interests, detours, goal displacement, goal and actor invention and concealing detours (see also Callon 1986a, b). By contrast, while it is clear that for Pierce et al. (2011) place frames are simplifications of the complexity of actors—wherein, for example, 'Bolivia's forests [become] modern, natural, sustainable and economically vibrant places' (p. 63)—the specific processes through which these simplifications might organize, and disorganize, actors within a place bundle, and dominate other place frames, remain unclear. Similarly, recent studies influenced by Pierce et al.'s (2011) framework also tend to under theorize the processes by which place simplifications are prescribed and negotiated (e.g. Foo et al. 2013; Pierce and Martin 2015; Rozema et al. 2015). For example in their study of a protest against a high-speed rail project, Rozema et al. (2015, p. 103) briefly mention how simplified placial images are constructed by the project sponsors and internalized by various actors including the protesters with important political ramifications, yet the specific organizational processes of prescribing/negotiating place simplifications are not discussed.

What is also notably missing from Pierce et al.'s (2011), and indeed Cresswell's (2015), relational reading of place, is the suggestion from ANT that place-making is not simply relational, it is regressive. More specifically, for one group of actors to enrol another in their place frame/simplification requires that they not only enrol a first order of actors—for example to persuade them 'I want what you want' (Latour 1987, pp. 109–110)—but consider the abilities and future capacities of those actors in turn to enrol a second order of actors in their own actor-networks by prescribing, even inventing, simplified identities and interests (Callon 1986a, b; Latour 1987). If any of these regressively nested actor-networks fails to cohere sufficiently, then it may follow they cannot readily accept the simplifications of their identity and interests prescribed by the original network builder. Considering place-framing in this way as part of an infinitely regressive not simply relational process of ordering, where actors enrol and are themselves enrolled (Hitchings 2003; Murdoch 1998), also positions place-framing alongside the framing of all manner of ontological entities, from forests, to markets, to firms. While Pierce et al. (2011) bracket off the work required to stabilize the objects that appear in place bundles (e.g. buildings, social groups, roads, parks, forests and corporations), ANT suggests that such embedded actor-networks may continue to play a role in the construction of other actors, including the identity and agency of network builders such as place producers and decision makers (Callon 1986a). By paying attention to the regressive, nested, character of processes of simplification, we can start to understand how complex and fragile place-making can be, and that not only does it create novelty as it gathers new actors together (Murdoch 1998)—the eventfulness of place-making (Casey 1998; Cresswell 2015)—but stabilization must work through such eventfulness. For instance, adapting Pierce et al.'s (2011) own illustrative example: even if a logging company within a Bolivian forest are enrolled to a simplified placial image of the forest as a sustainable global resource, rather than fragile habitat, and the forest assemblage (plants, streams, soil, policies, villages, roads, laws and norms) is modified accordingly, this reductive image cannot necessarily be defined in sufficient detail for it to prescribe in advance the behaviour of all actors enrolled by it (cf. Bear and Eden 2008; Murdoch 1998; Müller 2014). This material-semiotic excess, or overflow to framing (Callon 1999), suggests that at the very least ad hoc modifications, and enrolments, are required to sustain this ordering that may, in turn, start to generate emergent place bundles, new space-time trajectories for objects (Pierce and Martin 2015), and maybe new strategic place frames.

In this section, we have situated our rehabilitated ANT treatment of place, elaborated in the first two sections of this chapter, alongside recent approaches to relational place-making within human geography. While this review is necessarily limited, it serves the purpose of this chapter in identifying the extent to which ANT can indeed offer a *novel* way of understanding place. To this end, we have noted some strong similarities between ANT and relational approaches to place. That this is the case is perhaps not surprising given that many of these relational approaches are influenced by ANT and especially the work of Bruno Latour (e.g. Massey 2005; Pierce et al. 2011). However, it is notable that there remain significant possibilities for ANT to supplement work on relational place-making, in particular around the specific techniques through which places are ordered and stabilized for strategic ends, and moreover, how those ordering processes are not necessarily opposed to concepts of place as thrown together and eventful (Massey 2005; Pierce et al. 2011) but rather mutually constitutive. What is also particularly evident in relational approaches to place-making is not only the lack of theoretical insight into strategic processes of place-making but the absence of ANT notions of (infinite) regression which helps define those processes (Callon 1986a; Latour 1987). ANT's ontology is not simply relational, it is infinitely regressive, and it is this regressive aspect that offers a potentially novel way to understand the complexity and fragility of place-making. In the next section, we will turn to our empirical case study of a housing development, Garendon Park in Loughborough, in order to develop these possibilities to think place with ANT.

Thinking ANT and Place-Making at Garendon Park

Garendon Park is a 466-hectare site approved for a future housing development of 3200 homes on the periphery of the market town of Loughborough, located in the English East Midlands. The development, formally titled the West of Loughborough Sustainable Urban Extension, will increase the population of Loughborough by between 10 and 20%, as well as provide two new primary schools, a park for travellers, access to restored parkland, as well as space for business and retail units; development is planned to take place in stages by two housebuilding companies over a period from 2016/2017 to 2031/2032. Unsurprisingly, given its relative scale to the size of existing settlements, the project has attracted controversy in particular due to its potential to engulf two other smaller settlements (the town of Shepshed and village of Hathern), the presence of protected historical sites within the development boundaries, and doubts over the capacity of local transport infrastructure to cope with increased demand. These concerns galvanized a series of protests by nearby residents, which became organized into the Garendon Park Conservation and Protection Group (GPCPG). Despite these protests, the proposed development was approved at a Planning Committee meeting of Charnwood Borough Council on the 22nd September 2015.

As with many, if not all, housebuilding projects, at least in the Global North (compare with Grubbauer this volume), the Garendon Park development proposal and the ensuing planning decision process were pervaded by a series of interacting place-based simplifications, negotiations and stabilizations. The planning proposal documents and audio records of Planning Committee decision for Garendon Park are mobilized here to understand how multiple place simplifications are stabilized and mediated, drawing upon, and developing, concepts derived from ANT. Our starting point for the empirical analyses presented here is the notion, described above, that place-making is not simply relational, it is regressive, and it is this regressive dynamic which is key to how place simplifications are accepted, become stabilized and serve specific political ends and projects. But it is only through empirical analyses of such negotiations, as at sites such as Garendon Park that we can start to elaborate on the specific processes that constitute these regressive acts of place-making.

To understand how ANT can inform an analysis of the stabilization of place-making, it is important to first establish some of the key place simplifications at play surrounding the Garendon Park development. These simplifications, which are akin to the place frames described by Pierce et al. (2011), correspond to strategic attempts to simplify images of the complexity of groups of heterogeneous actors to produce a certain image of Garendon Park as a place with a particular past, present and future. As with Callon (1986a), these processes of simplification can be understood as processes of actor-network stabilization. While it is clear that many of these simplifications overlap, not least as they were mobilized and deliberated over within planning decision processes, they can be analytically separated. Within our analysis, we identified five specific place simplifications, each involving the enrolment of certain actors with their own attendant actor-networks and simplifications (Table 7.1).

In our case, two network builders can readily be identified as the GPCPG and the housing developers, each attempting to bring into being a very different bundle of spatial and temporal trajectories (Pierce et al. 2011) for the site that revolved around acceptance, or not, of the stabilization of place simplification #2. In short, the developers promoted the site for development, GPCPG rejected all development. In the remainder of this section, we will discuss four overlapping tactics through which network builders can effect enrolment, stabilize actors and, in this case, pursue a particular place simplification. Specifically, we will now consider the value of these techniques to understand how it was that place simplification #2 was stabilized and became indispensable to all the other place simplifications.

Table 7.1 Place simplifications and processes of enrolment at Garendon Park development

Place simplification	Description	Some key actors invoked
1. Garendon Park as irreplaceable farming land	This site contains high-quality farm-land and helps feed a growing population, protect biodiversity and mitigate flood-ing, especially as com-pared to other sites	Future growing population, arable crops, local farmers, farmland wildlife, Natural England (stat-utory environmental impact approval body), flooded water-courses and buildings, other 'greenfield' and 'brownfield' sites, GPCPG
2. Garendon Park as an ideal site for urban development	The population is growing and we need more houses and this site is very well-con-nected to local services compared to other possible sites	Future growing population, other towns and villages, planning strategies (local and national) that establish need for more housing, housing developers, councillors, planners, other 'greenfield' and 'brownfield' sites
3. Garendon Park as a 'green wedge'	Garendon park separates Hathern, Loughborough and Shepshed and pro-tects their distinctive identities	Hathern, Shepshed, Loughborough, future planning applications, GPCPG, housing developers, planners, councillors, planners
4. Garendon Park as a historical resource	The site is an important historical resource containing 14 desig-nated heritage assets, including the remains of a medieval abbey and a Grade I listed triumphal arch	Current historic sites, Historic England (statutory historic impact approval body), GPCPG, housing developers, councillors, planners
5. Garendon Park as a leisure space	This site is an impor-tant leisure resource for walkers, bikers and horse riders, containing rights of way and forming part of the National Cycle Network	Health and well-being of local res-idents, green spaces serving local residents not wider community, GPCPG, councillors, planners

Tactic One: Enrol More Black Boxes Than Your Opponent Can Open

The first tactic involves the identification and enrolment of 'black boxes'—actors whose interests, identities and arguments are widely accepted—that will lend support to your nascent actor-network (Callon 1986a, b; Latour 1987). In our case, two prominent network builders, the housing developers and GPCPG, both identified various 'black boxed' actors to stabilize place simplifications. For example in their Planning Statement, establishing the rationale for the development, the housing developers made repeated appeal to local authority's Core Strategy (Charnwood Local Plan 2015) and the National Planning Policy Framework (NPPF 2012) to justify the suitability of the site (simplification #2):

> The most recent version of the emerging Core Strategy is the Submission version which is at examination in 2014, and includes the Site as a proposed allocation in Policy CS22. The application proposals in this submission address the policy requirements of Policy CS22 comprehensively in respect of development requirements and infrastructure ... NPPF paragraph 49 confirms that housing applications should be considered in the context of the presumption in favour of sustainable development and states that relevant policies for the supply of housing should not be considered up to date, if the local planning authority cannot demonstrate a five year supply of deliverable housing sites. As set out in Section 7 of this Statement, [the local authority] cannot currently demonstrate a five year supply of deliverable housing and the 'presumption in favour of sustainable development' of the NPPF applies. (Planning Statement 2014, p. 6)

The Planning Statement reveals how the developers had worked closely with the local authority over a few years to develop the proposal documents as part of the local authority's own Core Strategy which itself was a response to national planning policies, especially the NPPF (2012). Significantly, the Core Strategy identifies Garendon Park as part of a West of Loughborough Growth Area to 'make a significant contribution to our strategic housing needs' (Charnwood Local Plan 2015). While the NPPF (2012) does not include specific housing targets, it does indicate that local authorities without specific targets and plans (i.e. an approved Core Strategy) are to approve housing developments if they adhere with the principles of the NPPF (2012) including consideration of larger scale development and urban extensions (NPPF 2012, pp. 13–14), thus lending further support to place simplification #2. The GPCPG adopted a similar strategy of recruiting 'black boxed' actors as it enrolled 287 local residents to write letters in support to stabilize place simplifications (#1, #3, #4 and #5) in order to reject the preference for development (place simplification #2). For instance, one resident identified the value of the agricultural land at Garendon Park, and evoked another 'black box'—the preference in the NPPF (NPPF 2012, p. 26) to develop only on poor quality agricultural land, to make the case against development at this site.

Each of these 'black boxed' actors was enrolled by a network builder, whether the housing developers or GPCPG, to lend their support towards the stabilization of a specific place simplification. Yet it is clear that the sheer number of 'black boxes' presented by the housing developers (NPPF, Core Strategy, pre-proposal consultations with the local authority, environmental surveys statements, traffic surveys and models, pre-proposal negotiations with statutory organizations such as Natural England) was far more extensive and was thus considerably more resource intensive to open up than the equivalent information presented by the GPCPG. As the GPCPG complained in their statement about the development of the Planning Committee, they were faced with an:

> … enormous amount of supporting documentation provided by the developers, [thus] it has not been possible to scrutinise every detail. It is for this reason that GPCPG believe that a much longer time frame than normal should be given for objections to be submitted. (GPCPG consultation letter 2014)

We can start to understand how the size of the effort of the housing developers to stabilize an actor-network around their favoured place simplification #2 had a distinct advantage due to the sheer number of different supportive 'black boxed' actors they could enrol. Local residents were equally concerned about their capacity to resist the pile of 'black boxes' they faced:

It is alarming and sad to find this kind of opportunistic development being pursued with such indecent haste in the face of an imminent and considered assessment of its soundness. I expect to hear that Charnwood Borough Council has protected the best interests of its citizens (and electorate) and refused this application. (Letter sent by resident during public consultation)

The challenge here was that '... if one wishes to question a fact or bypass an artefacts one might be confronted by so many black boxes that it would become an impossible task' (Latour 1987, p. 179).

By contrast, the GPCPG simply did not have the resources to conduct traffic or environmental surveys and models to challenge the findings of those prepared by the housing developers. Moreover, many of the 287 letters drawn in support of the GPCPG enrolled the same limited number of 'black boxed' actors to construct their arguments, often reproducing a bulleted list of points produced by the leaders of the GPCPG. The level of repetition by local protesters was problematic insofar as, under UK planning law, the *strength* of local opposition to a proposal is considered a non-material factor and thus is not a basis for refusal. That is to say, under the UK planning system, what mattered was not the number of black boxes of the same type but the number of different types, each requiring different resources and techniques to open and contest.

Nevertheless, tactic 1 is not without its limitations. In particular, in order to enrol other actors you have to become enrolled by them; this is the problem of regression that ANT usefully draws our attention towards (Callon 1986a). Specifically, the stabilization of place simplification #2 now hinged on the stabilization of the Core Strategy (during the proposal decision the Core Strategy was still to be approved by the Planning Inspector), or the stabilization of the relative importance of specific principles in the NPPF. In other words, a well-stacked pile of black boxes of different types can aid place stabilization but not prevent processes of disordering. This is also the point that Murdoch (1998) makes concerning the limitations of prescription in network building. For example, one letter from a resident protesting the development demonstrates how readily actors seeking to challenge a place simplification can regress from one strategically nested 'black box'—the Core Strategy and the Council—to another contained within it:

You fob us off with fanciful estimates of new business and jobs in the area – where does this come from? You must think we are stupid not to realise that the only reason you would even consider this proposal is another 3200 council tax payments which would amount to more than £5m in your coffers to fritter away as you please to artificially boost your egos, pacify your voters and your conscience. (Letter sent by resident during public consultation)

Tactic Two: Cut Off Actors from Alternative Ways Forward

In response to the potential of being enrolled into another capricious, and labyrinthine, actor-network, Latour (1987) proposes network builders, such as the housing developers, employ a second tactic to stabilize an actor-network: to cut off actors from alternative ways of pursuing their explicit interests. In the context of our case, this tactic can be partly witnessed in the decision taken by the local authority, and the housing developers, to reject other alternative sites for large-scale housing development. In July 2014, just three months before the submission of the Garendon Park proposal, planning approval was refused for development of almost a thousand houses on a site east of Loughborough in the village of Coates. While the housing developers, as network builders, were not involved in this decision, they surely benefitted from the reduction in the number of alternative ways forward for housing development in the locality. After all, many of the enrolled supporters of the GPCPG accepted the need for housing but continued to advance the case that alternative sites for development, such as the site in Coates, or brownfield sites within Loughborough town centre, had not been properly considered. Given the earlier decision by the local authority to refuse planning for the Coates site and the lack of alternative interest by other housing developers in other sites, Garendon Park readily becomes an obligatory passage point (Callon 1986b) to meet recognized needs for future housing. Indeed while the GPCPG wishes to challenge the suitability of Garendon Park for development, they cannot themselves create an alternative plan for housing development without the support and resources of alternative housing developers. The obvious limitation with this tactic is, as Latour (1987) suggests, that it is very difficult to cut actors off from *all* alternative ways for them to pursue their interests, especially perhaps when the preferred alternative requires far less effort: build nothing! That is, many of the actors supporting the GPCPG, and thus opposing the stabilization of place simplification #2, continued to question the need for any housing development in the local area:

> I have no understanding why building 3400 houses is a good idea, Loughborough already becomes gridlocked almost daily, 3400 houses brings a potential 6000 extra vehicles on an already congested road network. (Letter sent by resident during public consultation)

Tactic Three: Detour Actors' Goals

To further address these limitations, a third tactic was mobilized by the housing developers; this involves the network builder asking actors that do not share their explicit interests to support them in order to reach their own goals. Latour (1987) suggests that this detour tactic is only feasible if the detour is short enough so that actors can be confident their interests and goals will still be served. Within our case, it is clear that the housing developers had already positioned the stabilization of place simplification #2 as a detour to stabilize place simplifications #1, #3, #4 and #5. Consider, for example, this extract from the Planning Statement:

> At present there is no public access to the Garendon Park. The proposals seek to enhance public accessibility and the permeability of the Site including opening up public access to the Park. This will greatly enhance the opportunities for recreation and leisure for both residents and visitors, as well as an appreciation of the value of the heritage assets … the Site has the potential to create a connected urban system of Loughborough and Shepshed, served by Garendon Park at the centre. This will benefit both Loughborough and Shepshed, whilst maintaining the physical and cultural identities of the towns. The Site is, however, physically separated

from Shepshed by the M1 Motorway. It is intended that the Development will maintain visual separation from Shepshed through sensitive landscaping and woodland buffer areas along its western boundary, whilst providing sustainable linkages. (Planning Statement 2014, pp. 4, 5)

Here, the place simplifications of heritage, leisure and distinctive identities (#4, #5 and #2 in Table 7.1), and thus the interests of supporters of the GPCPG, are given a shortcut to their stabilization by taking a detour via the stabilization of the place simplification of development (#2). What is notable is that these processes of enrolment are future-oriented, they depend on the construction of 'black boxed' future actors (new access paths, car parks, fencing, visitor centres, visitors to the parkland, woodland and landscaping) to mediate a detour rather than the enrolment of known actors and their explicit interests (as with Latour 1987; Callon 1986a). As these 'black boxes' are highly promissory rather than tangible, it is even more difficult for the GPCPG to open them up and challenge them. Nevertheless, as Latour (1987) recognizes, this tactic has its own limitations: What happens if it is not clear that the usual route of actors is cut off (i.e. tactic 2). In other words, the GPCPG could easily object that their favoured place simplifications, especially #1, #3 and #4, relating to the agricultural, 'green wedge' and heritage value of the site, are already stabilized by a range of actors (enforced planning policies and laws, fences, land ownership agreements and protection groups) and that, as such, no detour via the development of Garendon Park is required.

Tactic 4: Curtailment of Explicit Interests

To overcome the limitations of tactics 1, 2, and particularly 3, Latour (1987, pp. 113–199) presents a radical last solution to achieve actor-network stabilization: the abandonment of the explicit interests of actors. This tactic corresponds to a process of invention where problems are created to generate new explicit goals for actors (see also Callon 1986a, b). In our case, a particularly significant problem was evoked during the Planning Committee meeting by one elected councillor:

The other thing is, with the ringing endorsement we've had from the Government Inspector, I think it's inevitable that if we turn this down it will go to appeal and the appeal will be upheld. I think that's as sure as God made it ... It's incumbent upon us as a Council to ensure that this place becomes a place worth living in and what we should do is go through everything with a fine tooth comb before the building starts, look at the quality of life, the community centres, the medical centres, the quality of the schools, commuter roads, safe roads, bungalows for the elderly, it should be a gem in the crown really. (Charnwood Borough Council, Planning Committee 22nd September 2015)

While tactic 1 is clearly evident here (the 'black boxed' Government Inspector capable of acting from afar yet incapable of being challenged), the Councillor also presents a new problem to those considering refusing the development—the need to ensure that sufficient time and resources are available before construction to ensure the development takes into account place simplifications #1, #3, #4 and #5. If the proposal is rejected, then the local authority and community will, it is now claimed, then have less time, given the projected future decision of the Government Inspector, to ensure that the development will have addressed their interests. As Latour (1987) suggests, such problems are future orientated and often speculative, yet they can be powerful means of shifting the goals of actors. It is telling that the discussion after this juncture in the Planning Committee meeting entirely concerned the conditions which are to be attached to any approved planning decision for development, before the Committee vote result was announced 10-2 in favour of approving the development.

The Contribution of Housebuilding to ANT Approaches to Place-Making: Infinite Regression, Futures and Affect

Our ANT-informed analysis of the early stages of the Garendon Park housebuilding project indicates the contribution ANT studies of the construction industry, in our case housing development, can make to understanding place-making. Specifically, the four tactics we presented here, all derived from classic work within ANT (Callon 1986a, b; Latour 1987), indicate the value of ANT to understand processes through which specific place frames (Pierce et al. 2011) are stabilized. In particular, ANT suggests how the issue of regression is an important, yet neglected, aspect of place-making. Each tactic corresponded with how attempts to stabilize a place simplification are bound up with the arrest of (infinitely) regressive potentials in an actor-network. Tactic 1 concerns how this was effected through the sheer quantity of anteced-ent 'black boxes' being enrolled to support a place simplification ('black box +n'), especially important here is the ability to exceed a threshold where the number of black boxes exceeds the resources of your oppo-nents to open them up. Tactic 2 concerned the forestalling of alterna-tive actor-networks, and their black boxes, which may have led to the opening up of a regressive movement within the developers' actor-network ('black box -n'). Tactic 3 involved the potential promise of future actors produced to enrol recalcitrant existing actors in a specific place simplification ('the black box yet to come that will enrol your interests'). Finally, tactic 4 worked upon cultivating a feeling of risk as a problem which had to be addressed by accepting the 'black boxed' decision to develop ('the black box yet to come that will change your interests'). These four tactics were intended to arrest the very real potential for infinitely regressive movements by the GPCPG to open up the embed-ded actor-networks nested within place simplification #2. As ANT thinking suggests while 'In theory reality is infinite. In practice, as a result of the translations that it brings about, an actor-world is limited to a series of discrete entities whose characteristics or attributes are well defined' (Callon 1986a, p. 28). The forestalling of such infinite regres-sive movements, which is concomitant with the enrolment of actors to a support a place simplification, was essential for the developers to stabi-lize Garendon Park before and during the Planning Committee meeting as a suitable place for development.

Importantly, the stabilization processes we documented here, and the potential for infinite regressions, suggest that places, and perhaps all actors, are not simply the product of the alignment of heteroge-neous actors in the present but rather they are also dependent upon specific tactics intended to amplify and arrest future *potentialities*. This temporal dynamic is important because ANT thinking is often criticized, especially as contrasted to other relational approaches, such as assemblage and affect theories, for failing to develop a sufficient account of virtual potentialities. Instead, it is said to be wedded to an actual-ism, such that the unexpected future, the event, the aleatory, is down-played (cf. Knox et al. 2016; Müller and Schurr 2016; Thrift 2000; Whittle and Spicer 2008). In contrast, our analysis suggests that ANT can be developed through empirical studies to help us register the com-plex role of virtual potentialities, or affect—'a body's *capacity* to affect and to be affected' (Seigworth and Gregg 2010, p. 2)—in the organiza-tion of collectives of human and non-human bodies (Anderson 2014; Massumi 2002). The recognition of the virtual alongside the actual is central to concepts of affect. As Massumi puts it: 'affect is the virtual as point of view' (2002, p. 35)—"the 'where we might be able to go and what we might be able to do' in every present situation" (2015, p. 3; see also Anderson 2014, p. 10). Affects can be negative, like sadness or exhaustion, diminishing bodily capacities to act differently, or positive like joy or anger, augmenting future bodily capacities to act anew. Like any body, or in ANT terms 'actor', Garendon Park can be considered to be a set of actualized relations between different concrete components, such as trees, grass, paths, buildings, monuments, planning policies, each corresponding with a specific place simplification. But, the site is also a set of *potentials* for each of those components to enter into new relations with new actors in as yet unknown ways, creating capacities for the place itself, and perhaps those constituent actors, to *become* different (Anderson 2014; Massumi 2002).

Our case suggests two novel insights into how we can think about the eventful affectivity of place with ANT. First, the *potential* that an actor-network can collapse into an infinite regression of its nested actor-networks (Callon 1986a; Latour 1987) fosters both open-ended anxiety in those looking to stabilize that network, as well as open-ended hope in those looking to disrupt it. These two affective conditions ref-erence contrasting desires that function to act as driving forces to guide processes of network stabilization (cf. Müller and Schurr 2016). In other words, if the actors enrolled to constitute a place simplification (build-ings, people, roads, stories, cultural practices, soil, weather, planning pol-icies and Government Inspectors) no longer possess a *potentiality* to be identified as discrete entities—that is, if the problem of infinite regress is not arrested—the potential for stabilizing that place may also recede from view. The amplification of affectivity in place simplification #2, the opening up of the virtual in the 'black boxes' as enrolled by the developers (e.g. questioning the presumed coherence of the NPPF to support the application by discussing the quality of agricultural land to be lost), was precisely the tactic employed by the GPCPG to encourage the local authority to reject the proposed housing development. Second, virtual potentialities also animated all the network building tactics of stabilization intended to arrest this very potential for infinite regression—whether the 'black box +n' (tactic 1), the 'black box -n' (tactic 2), 'the black box yet to come that will enrol your interests' (tactic 3) and 'the black box yet to come that will change your interests' (tactic 4). Each of the tactics employed by the housing developers also works with potentiality—whether the decision yet to come of the Planning Inspector, or the potential installation of hedgerows. These negotiations did not refer to actually known abilities of actors to relate together but to future capacities for actors to be affected and affected differently in the future in order to form emergent becomings (Massumi 2015) of Garendon Park.

The intensity of these movements of affect was, as Massumi (2002) suggests, highly evident in bodily registers too, as evidenced in the depth of transitions in affective states as individuals felt more or less able to act in the future (Anderson 2014; Massumi 2015). We might consider, for example: the shock, sadness, alarm and occasional hope (of planning refusal) expressed in letters objecting to the development; the loud applause and augmentation of protesters' agency that followed those speaking at the Planning Committee meeting to oppose the development; the silence and stillness of diminished agency after the vote was taken in favour of the development. The ebb and flow of these affective intensities, which is beyond the analytical scope of this chapter, all document how ANT can help develop the significance of the virtual, potentiality and affect in the stabilization of place.

Housebuilding, and indeed construction, does not represent the only empirical opportunity to develop the potential for ANT to understand relational place-making (Cresswell 2015; Pierce et al. 2011), nor indeed to develop the role of virtual, potentiality, and affect with ANT (Müller and Schurr 2016). Nevertheless, housebuilding does provide a particularly rich opportunity to witness how place simplifications

become stabilized. And, in particular, how affective futures and potentialities produced through these place-making processes interact to
produce stabilized outcomes at events such as Planning Committees
meetings. Of course, while the planning and design phase of Garendon
Park represents a vital moment in the development of this housing project, the developers must continue to enrol and align actors they now
appear, on paper at least, to have organized (Callon 1986b; Latour
1987). New actors will need to be incorporated and enrolled during
the construction phase—bricklayers, soils, planning conditions, wildlife, building control officers and customers—and at each new bodily encounter, new tactics for enrolment must be deployed, and new
affects, or potentials for agency, will emerge that can undo but also render possible processes of place-making.

References

Allen, J. (2011a). Topological twists: Power's shifting geographies. *Dialogues in
Human Geography, 1*(3), 283–298.

Allen, J. (2011b). On actor-network theory and landscape. *Area, 43*(3),
274–280.

Amin, A. (2002). Spatialities of globalisation. *Environment and Planning A,
34*, 385–399.

Anderson, B. (2014). *Encountering affect: Capacities, apparatuses, conditions.*
Farnham: Ashgate.

Antonisch, M. (2011). Grounding theories of place and globalisation.
Tijdschrift voor Economische en Sociale Geografie, 102(3), 331–345.

Bear, C., & Eden, S. (2008). Making space for fish: The regional, network and
fluid spaces of fisheries certification. *Social and Cultural Geography, 9*(5),
487–504.

Building for Life. (2015). *Building for life 12: The sign of good place to live,*
Building for Life Partnership. Retrieved from http://www.designcouncil.
org.uk/sites/default/files/asset/document/Building%20for%20Life%20
12_0.pdf.

Callon, M. (1986a). The sociology of an actor-network. In M. Callon,
J. Law, & A. Rip (Eds.), *Mapping the dynamics of science and technology*
(pp. 19–34). Basingstoke: The Macmillan Press.

Callon, M. (1986b). Some elements of a sociology of translation: Domestication of the scallops and the fishermen of St Brieuc Bay. In J. Law (Ed.), *Power, action and belief: A new sociology of knowledge?* (pp. 196–223). London: Routledge.

Callon, M. (1991). Techno-economic networks and irreversibility. In J. Law (Ed.), *A sociology of monsters: Essays on power, technology and domination* (pp. 132–164). London: Routledge.

Callon, M. (1999). Actor-network theory—The market test. In J. Law & J. Hassard (Eds.), *Actor network theory and after* (pp. 181–195). Oxford: Blackwell.

Casey, E. (1998). *The fate of place: A philosophical history*. Berkeley: University of California Press.

Castells, M. (2000). *The rise of the networked society*. Oxford: Wiley-Blackwell.

Charnwood Local Plan. (2015). *Charnwood Local Plan 2011 to 2028*. Core Strategy Adopted 9 November 2015. Retrieved from http://goo.gl/KNLL2l.

Collins, H., & Yearley, S. (1992). Epistemological chicken. In A. Pickering (Ed.), *Science as practice and culture* (pp. 301–326). Chicago: Chicago University Press.

Cresswell, T. (2015). *Place: An introduction*. Oxford: Wiley-Blackwell.

Ettlinger, N. (2003). Cultural economic geography and a relational microspace approach to trusts, rationalities, networks and change in collaborative workplaces. *Journal of Economic Geography, 3*(2), 145–171.

Foo, K., Martin, D., Wool, C., & Polsky, C. (2013). The production of urban vacant land: Relational placemaking in Boston, MA neighborhoods. *Cities, 35,* 156–163.

Gadd, K. (2016). Street children's lives and actor-networks. *Children's Geographies, 14*(3), 295–309.

Graham, S. (1998). The end of geography or the explosion of place? Conceptualizing space, place and information technology. *Progress in Human Geography, 22*(2), 165–185.

Hitchings, R. (2003). People, plants and performance: On actor network theory and the material pleasures of the private garden. *Social and Cultural Geography, 4*(1), 99–114.

Jones, A. (2008). Beyond embeddedness: Economic practices and the invisible dimensions of transnational business activity. *Progress in Human Geography, 32*(1), 71–88.

Kavaratzis, M., & Ashworth, G. (2005). City branding: An effective assertion of identity or a transitory marketing trick? *Tijdschrift voor Economische en Sociale Geografie, 96*(5), 506–514.

Knox, H., O'Doherty, D., Vurdubakis, T., & Westrup, C. (2016). Something happened: Spectres of organization/disorganization at the airport. *Human Relations, 68*(6), 1001–1020.

Kriese, U., & Scholoz, R. (2012). Lifestyle ideas of house builders and housing investors. *Housing, Theory and Society, 29*(3), 288–320.

Latour, B. (1987). *Science in action: How to follow scientists and engineers through society*. Cambridge, MA: Harvard University Press.

Latour, B. (1999). On recalling ANT. In J. Law & J. Hassard (Eds.), *Actor network theory and after* (pp. 15–25). Oxford: Blackwell.

Latour, B. (2005). *Reassembling the social: An introduction to actor-network Theory*. Oxford: Oxford University Press.

Law, J. (1986). On the methods of long distance control: Vessels, navigation, and the Portuguese route to India. In J. Law (Ed.), *Power, action and belief: A new sociology of knowledge? Sociological Review Monograph 32* (pp. 234–263). Henley: Routledge.

Law, J. (2002a). Objects and space. *Theory Culture Society, 19*(5/6), 91–105.

Law, J. (2002b). *Aircraft stories: Decentering the object in technoscience*. Durham, NC: Duke University Press.

Lee, N., & Brown, S. (1994). Otherness and the actor-network: The discovered continent. *American Behavioral Scientist, 37*(6), 772–790.

Lees, L., & Baxter, R. (2011). A 'building event' of fear: Thinking through the geography of architecture. *Social and Cultural Geography, 12*(2), 107–122.

Malpas, J. (1999). *Place and experience: A philosophical topography*. Cambridge: Cambridge University Press.

Massey, D. (1991, June). A global sense of place. *Marxism Today*, 24–29.

Massey, D. (2005). *For space*. London: Sage.

Massumi, B. (2002). *Parables for the virtual: Movement, affect, sensation*. Durham, NC: Duke University Press.

Massumi, B. (2015). *Politics of affect*. Cambridge: Polity Press.

Müller, M. (2014). The topological multiplicities of power: The limits of governing the Olympics. *Economic Geography, 90*(3), 321–339.

Müller, M. (2015). A half-hearted romance? A diagnosis and agenda for the relationship between economic geography and actor-network theory (ANT). *Progress in Human Geography, 39*(1), 65–86.

Müller, M., & Schurr, C. (2016). Assemblage thinking and actor-network theory: Conjunctions, disjunctions, cross-fertilisations. *Transactions of the Institute of British Geographers, 41*(3), 217–229.

Murdoch, J. (1998). The spaces of actor-network theory. *Geoforum, 29*(4), 357–374.

Mutch, A. (2002). Actors and networks or agents and structures: Towards a realist view of information systems. *Organization, 9*(3), 477–496.

Nicol, C., & Hooper, A. (1999). Contemporary change and the housebuilding industry: Concentration and standardisation in production. *Housing Studies, 14*(1), 57–76.

NPPF. (2012). *National Planning Policy Framework, Department of Communities and Local Government*, March 2012. Retrieved from https://goo.gl/pYlqTv.

Pierce, J., & Martin, D. (2015). Placing Lefevbre. *Antipode, 47*(5), 1279–1299.

Pierce, J., Martin, D., & Murphy, J. (2011). Relational place-making: The networked politics of place. *Transactions of the Institute of the British Geographers, 36*(1), 54–70.

Planning Statement. (2014). *West of Loughborough sustainable urban extension planning statement.* William Davis Ltd. and Persimmon Homes PLC. Available from Charnwood Borough Council Planning Explorer. Retrieved from https://portal.charnwood.gov.uk/Northgate/PlanningExplorerAA/Home.aspx.

Pred, A. (1984). Place as historically contingent process: Structuration and the time-geography of becoming places. *Annals of the Association of American Geographers, 74*(2), 279–297.

Rozema, J., Cashmore, M., Bond, A., & Chilvers, J. (2015). Respatialization and local protest strategy formation: Investigating high-speed rail megaproject development in the UK. *Geoforum, 59*, 98–108.

Sack, R. (1997). *Homo geographicus.* Baltimore: John Hopkins University Press.

Seigworth, G., & Gregg, M. (2010). An inventory of shimmers. In M. Gregg & G. Seigworth (Eds.), *The affect theory reader* (pp. 1–28). Durham, NC: Duke University Press.

Smith, R. (2003). World city actor-networks. *Progress in Human Geography, 27*(1), 25–44.

Star, S. (1991). Power, technology and the phenomenology of conventions: On being allergic to onions. In J. Law (Ed.), *A sociology of monsters: Essays on power, technology and domination* (pp. 26–56). London: Routledge.

Thrift, N. (1996). *Spatial formations.* London: Sage.

Thrift, N. (2000). Afterwords. *Environment and Planning D: Society and Space, 18*, 213–255.

Whittle A., & Spicer, A. (2008). Is actor network theory critique? *Organization Studies, 29*(4), 611–629.

Woods, M. (1998). Researching rural conflicts: Hunting, local politics, and actor-networks. *Journal of Rural Studies, 14*(3), 321–340.

Young, N. (2010). Globalization from the edge: A framework for understanding how small and medium-sized firms in the periphery 'go global'. *Environment and Planning A, 42*, 838–855.

8

Organizing Space and Time Through Relational Human–Animal Boundary Work: Exclusion, Invitation and Disturbance

Daniel J. Sage, Lise Justesen, Andrew Dainty, Kjell Tryggestad and Jan Mouritsen

Introduction

Unruly individuals (Parker 2014), technologies (Winiecki 2009) and their attendant spacings and timings (Fleming and Spicer 2004; Quattrone and Hooper 2005) are readily shown to be effaced (Cederström and Fleming 2012), removed (Clegg et al. 2006, p. 17; Parker 2014) or discarded by human managers. Such troublesome

---

This chapter was originally published under the same title in *Organization*, reprinted here with permission of the publisher, Sage (http://journals.sagepub.com/doi/full/10.1177/1350508416629449).

D. J. Sage (✉)
School of Business and Economics, Loughborough University, Loughborough, UK
e-mail: d.j.sage@lboro.ac.uk

L. Justesen · K. Tryggestad
Department of Organization, Copenhagen Business School, Copenhagen, Denmark

actors, and their spacings and timings, thus appear to be both a target and limit for managerial domestication. Accordingly, in order to better understand the significance of boundary work in human organizing, and especially management, seems to require us to confront situations where 'wild' actors object to their domestication. Thus, it might be expected that animals—the archetypical target, and perhaps also limit, of human domestication—present an exemplary opportunity to study the significance of boundaries in human organizing. However, this task is not unproblematic. Beyond organization studies, animals are frequently presented as being controlled *by* and *for* management: underweight, and unproductive, salmon are destroyed in fish farms (Law and Lien 2013); mating, and obscene, chimpanzees removed from public view in zoos (Grazian 2012); infective bacteria cleaned from poultry sheds (Miele 2011); uncooperative cows taken from the automatically milked herd (Holloway 2007); and younger, more profitable, alligators sifted into skin farms (Keul 2013). These studies thus offer compelling stories of domestication where animals become controlled by managers to serve, or not serve, organizational ends, entering into or out of preformed organizational spaces and times.

The problem is that if animals only appear as significant in relation to human organizing as passive 'objects' that comply, or not, with their managed domestication, it is difficult to understand the multiple ways their 'wildness' makes a difference to, rather than is placed outside, human organizing. This is the starting point for our analysis: to challenge the boundary work that currently places animals outside of processes of organization. By drawing on Actor–Network theory (ANT) (Latour 2005), we explore somewhat paradoxical possibilities that human organizational boundary work, even that which appears to

A. Dainty
School of Architecture, Civil and Building Engineering,
Loughborough University, Loughborough, UK

J. Mouritsen
Department of Operations Management,
Copenhagen Business School, Copenhagen, Denmark

exclude animals from human organizing, can be conceptualized as the relational outcome of complex entanglements *between* human–animal agencies, spacings and timings. Thus, the aim of our paper is to problematize the urge towards domestication that places animals outside human processes of organizing, and indeed management. As such, we also aim to contribute towards wider debates that discuss the role of 'wild', disorganizing agencies, spacings and timings, alongside the boundary work constitutive of human processes of organizing, and in particular those of management (Bloomfield and Vurdubakis 1999; Clegg et al. 2005, 2006; Knox et al. 2015).

This argument for (re)thinking organization with animals brings together different strands of thinking in and around ANT on: relational approaches to organization (Latour 2013); animal geographies (Bear and Elden 2011; Holloway 2007; Philo and Wilbert 2000); heterogeneous agencies (Callon 1986; Latour 2004, 2005; Law and Mol 2008); and relational spacings and timings (Latour 1986; Law 2002a). Our argument is worked through two illustrative empirical cases of human–animal organizing in the construction industry. Construction is not unique in contributing novel insights into the significance of animals within human organizational life, but the managed production of the built environment presents us with a set of organizing practices that are visibly involved in producing separate human and animal spaces and times (Philo and Wilbert 2000). It thus offers an important setting to investigate relations between human and animal boundary work.

The paper is arranged into three sections. First, we critically examine the boundary work that locates animals outside of human organizing. Then, second, in order to investigate animals as potentially making a difference to such acts, we turn to relational approaches to organization influenced by various ANT studies, and animal geographies, to recognize that *animal* actors can be involved in organizing. Analytically, this involves a triad of concepts—*Invitation, Exclusion* and *Disturbance*—which carry to the fore particular practices of relational boundary work between humans and animals. These boundary concepts are understood relationally, drawing attention to how capacities for humans to invite, exclude and disturb animals are entangled with capacities for animals to invite, exclude and disturb humans. Third, we draw upon two empirical studies of construction projects to illustrate how animals make a difference to organizing and management. Drawing on *Invitation, Exclusion* and *Disturbance*, this investigation provides insights into the ways in which organizational, and management, boundary work can be enacted *with* and *through*, rather than simply *against* and *beyond*, animal agencies, spacings and timings.

Organizing and Boundary Work

Organizational theories tend to locate animals outside the boundaries of organizing in two principle ways. The first approach, which we term 'managerialist', draws from organizational theories such as transaction cost economics, resource-based views of firms, systems theories and other related approaches (Santos and Eisenhardt 2005; Schreyögg and Sydow 2010). In their well-cited review of organizational boundaries, Santos and Eisenhardt (2005) explain that for organizations to survive and thrive, a select group of human managers must do boundary work, via control techniques, to delineate an organization's *efficiency, power, competency* and *identity* in a volatile global market environment. Within such accounts, 'environment' quickly becomes a synonym for 'business environment' (Klikauer 2013, p. 207), just as 'organizing' equates to 'management' (Parker 2002, p. 200). Such approaches are said to rehearse a teleology whereby 'management is a precondition for an organized society' and 'social progress is equivalent to our ability as human beings to increasingly control the nature worlds around us' such that 'Where we were once the victims of a wild, unruly nature, we are now becoming masters' (Parker 2002, pp. 2–3; cf. Klikauer 2013, p. 41). This managerialist discourse denies non-human agencies—including animals—a role in organizing unless they passively serve as a marketable resource for market managerial ends, as in 'green' accountancy (Milne and Gray 2013) and planetary geo-engineering (Anderson 2014). Thus, 'management' is to be defined not only as the naturalization of elitist hierarchies, generalized techniques of control or the primacy of maximizing shareholder value, but also as the commodification of 'Nature' (Klikauer 2013; Newton 2009; Parker 2002)—the exclusion of animals from acts of organizing.

A parallel approach to the evisceration of animals from organizing can be found in some poststructuralist accounts, especially those influenced by Cooper's *Organization/Disorganization* (1986). As Cooper (1986) explains, drawing on Derridian deconstruction, 'In its most fundamental sense, organization is the appropriation of order out of disorder' and so 'social structures are ordered, organized and made decidable always at the boundary line between opposing forces, between inside and outside, good and bad' (p. 327). For Cooper (1986), the ordering and decidability within organizations is thus defined through a process of 'Othering' where monstrous excesses of disorder and undecidability are purged: 'The struggle for the "superior" position necessarily requires the "support" of an "inferior" positions inasmuch as the latter is what defines the former' (p. 328); it is that 'the purging of 'bad' which threatens the system's internal purity and security' (p. 327). Unruly and wild agencies, including animals, are thus prefigured as productive negations that define organization in terms of order, decidability and control (Clegg et al. 2005, pp. 152–153). Inspired by Cooper (1986), Bloomfield and Vurdubakis (1999), for instance, discuss how the UK's Ministry of Deference sought to 'Other' the monstrous, disorganizing, agency of the possibility of a big cat killing sheep on Bodmin Moor. These, and other organizational scholars (e.g. Riach and Kelly 2015; Thanem 2006), have shown how organizing is enacted through boundary work that involves the negation of monstrous, disorganizing 'Others'. No matter how much we might celebrate the radical disruptiveness of monsters to disorganize (as in Riach and Kelly 2015; Thanem 2006), 'monsters', including animal agencies, are inevitably reproduced in such accounts beyond the boundaries of socially valued ways of organizing and talking about organizations.

A rather different take on the significance of boundaries to processes of organizing and therein the role of animals can be found in the work of Latour (2013). His work parallels some recent organizational theory (e.g. Clegg et al. 2005; Knox et al. 2015) in amalgamating notions of organization/disorganization: 'To organize is not, cannot be, the opposite of disorganizing. To organize is to pick up, along the way and on the fly, scripts with staggered outcomes that are going to *disorganize* others' (Latour 2013, p. 393, emphasis in original). For Latour, the significance of boundaries to organization does not concern generalized organization/disorganization distinctions, but rather the coordination of 'contradictory injunctions' (Latour 2013, p. 396) that limit us spatially, temporally and agentially (Latour 2013, p. 397), so as to effect 'something that has borders, frontiers, mandates, limits, walls, ends' (Latour 2013, p. 398). This 'something' is a temporarily stable organization (e.g. a regular meeting, corporation or enduring empire). To negotiate these contradictory injunctions, and their actual and potential spatial, temporal and agential limitations, actors are said to perform spatial, temporal and agential boundary work to do, undo and transform the limitations and contradictions placed upon them and other actors (Latour 2013). Boundaries are thus constituted relationally *within* not on the margins of acts of organizing, as actors spatially, temporally and agentially organize and disorganize themselves and others. Thus, crucially for our purposes here, Latour also suggests that actors are themselves relational effects (Latour 2005); as such, non-humans, including animals, can be considered actors if they make a difference to these spatial, temporal and agential processes of organizing.

To be clear, Latour does not suggest that non-humans, including animals, act *sui generis* to set boundaries on humans, that is, to dis/organize us; indeed, Latour (2005) rejects the idea that agency is a distinctly human or non-human property. Rather, as per ANT, agency is *relational*, based on differences that recursively produce more differences; it is constituted by the interplay of '*different* types of forces … because they are *different*' (Latour 2005, p. 75, original emphasis). ANT thus reverses any notion that agency is the basis of difference, say between humans and non-humans (contra Collins and Yearly 1992; Whittle and Spicer 2008); instead, the intermingling differing of difference produces agency. Viewed in this way, agency is disentangled from (human) intentionality (Law and Mol 2008, p. 58): an actor can be anything that makes a difference to the situation at hand—a '*moment of indeterminacy* that generates events and situations' (Law and Mol 2008, p. 74, original emphasis). Agency, including that involved in the boundary work that constitutes organization (as described by Latour 2013), thus is itself constituted through recursively transformative interactions between human and non-human actors.

In short, ANT draws attention towards 'how places and entities are "framed" by actions/agencies from elsewhere' (Hassard et al. 2008, p. 63). ANT's relational approach to boundaries thus also demands that rather than 'space' and 'time' pre-existing their relations, focus is given on 'how different spaces and times may be produced inside the networks built to mobilise, cumulate and recombine the world' (Latour 1986, p. 228). Elaborating on this point, Latour (1986, p. 230) offers the example of early European maritime expeditions: new ships connected European capitals to distant peoples, goods and lands, broadening spatial horizons, and creating new times that interrupted the routines of the past. Space-time is herein pleated and folded (Murdoch 1998) because distant objects such as foreign coastlines or goods circulate inside nearby stabilized networks, such as those formed between ships, states, water, stars, wind, compasses and various other navigational instruments (Law 2002a; Latour 1986). These extended, stabilized networks permit certain objects, such as ships or goods, to travel without deformation just as the extension of similarly stabilized networks allows scientists to act as spokespeople for certain 'universal facts' no matter when and where they travel (Callon 1986; Latour 1986). Viewed as such, the relational boundary work through which bounded actors and agencies are constituted also delineates spacings and timings. Any particular, delimited, space-time thus becomes a product of drawing alignments between certain actors (see Jones et al. 2004; November et al. 2010). Despite important debates by ANT proponents on this subject, for example around the ontological primacy of networks as constitutive of space (compare, e.g., Latour 1986 and Law 2002a), ANT steadfastly suggests the importance of *relations* in constituting boundaries of agency, space and time. This is why Mol and Law (2005) suggest of ANT: 'the problematisation of boundedness does not mean the end of boundaries' (p. 641). However, thus far this body of work has remained rather detached from theories of *organizational* boundary work (including that of Latour 2013). To understand its significance to such debates, we now turn towards ANT-influenced animal geographies.

Boundary Work and Animals

ANT research has long inspired animal geographies, extending early ANT studies of relational ordering that included animals (e.g. Callon 1986). These studies focus on the specific differences animals make to boundary work; they also offer points of reference to analyse how animals make a difference to organizing. Holloway (2007), for instance, examines how dairy cows interact with automated milking machines in complex ways, requiring spatial and temporal modifications and adjustments to the barn in order to promote successful milking (e.g. placing non-return gates between the automated milking machine and food and water, allowing lower ranked cows access to leave the milking barn if they are threatened by a higher ranked cow). As he explains: 'technology and layout [of the barn] are affected by the cows' bodies and behaviours, and bodies and behaviours are affected by the technology and layout' (Holloway 2007, p. 1051). Here, animals are shown to make a difference to how humans, presumably farm managers, organize as they establish boundaries around the spacings, timings and agencies of cows to pursue (cost) efficiencies in dairy farming.

Other animal geographies help us understand how animals and humans might organize, that is, set agential, spatial and temporal limits (Latour 2013), in more fluid ways (and more fluid space-times—see Law 2002a)—for example Bear and Eden's (2011) study of angling. It shows how in their riverine interactions, anglers and fish exchange certain agential capabilities, as they adapt to, and to an extent modify each other's bodily rhythms (rephrasing Deleuze and Guattari 1987—anglers 'become-fish' and fish 'become-human'). For instance, they write how 'their [the anglers] inability to see below the surface of the river … is often balanced out by honing their other senses, such as touch (in feeling the shape of the riverbed by dragging a weight across on the end of a fishing line)' (Bear and Elden 2011, p. 346); this practice is analogous to the barbles used by riverine fish in murky water to find their food. Bear and Elden (2011) also discuss how some fish avoid feeding in those times of the days most frequented by anglers. Thus, the agencies, spacings and timings of humans and animals become entangled, but only in order to reinforce boundaries more forcibly: for humans to minimize the time spent between catches and for animals to render their more fluid space alien to humans and evade being caught. Instead of deriding anthropomorphisms, like Whittle and Spicer (2008), animal geographers readily propose a 'guarded anthropomorphism' (Philo and Wilbert 2000, p. 20) as an ethical virtue to register our shared agency, spacings and timings, our shared suffering and delight (Greenhough and Roe 2011), while not downplaying relationally given differences, detachments (Ginn 2014) and boundaries (Bear and Eden 2008).

Animal geographers propose various ways in which animal agencies inflect human processes of organizing. However, the boundary work they describe often appears somewhat domesticated inside a set of stabilized human organizations (e.g. farms, laboratories and supermarkets) wherein spatial, temporal and agential boundaries are determined by human managers to pursue (cost) efficiency in the market (e.g. Greenhough and Roe 2011; Holloway 2007; Miele 2011). Animals that refuse to accept their institutionalized domestication are then simply systematically removed by management (Holloway 2007). Alternatively, when more fluid, uncertain, agential interactions between animals and humans are described, these tend to be positioned against (Keul 2013), or beyond (Bear and Elden 2011; Ginn 2014), the domain of management. Consequently, the variety of ways animals might make a difference to how humans organize, and especially manage, remains neglected. The following empirical study is intended to provide a more symmetrical account that registers the different ways humans organize, even manage, *with* animals.

Cases and Methodology: Building Organizations with Animals

Contemporary construction represents one of the most important, yet remarkably overlooked, places to study human/animal encounters (Sage et al. 2014). The construction industry is a major threat to animals and their habitats spanning a wide continuum from the temporary displacement of wildlife by noise and dust during animal extinctions and the permanent destruction of entire ecosystems. The construction industry has also become a focus for innovative wildlife experiments, from small-scale urban conservation projects (Lorimer 2008) to green infrastructure megaprojects (Tajima 2003).

Our two cases in infrastructure and house building help to illustrate the entanglement of animal and human boundary work rather than offering detailed, empirical case studies. While sharing a common project-based organizational form, and other similar organizational contexts (despite one being located in the UK and one in Scandinavia), the two cases illustrate different boundary work practices wherein animals, and their habitats, are involved in organizing. To help focus on relational boundary work, we delineate a set of analytical concepts which explain how boundaries keep actors apart, how boundaries foster new collective agencies, and how the dynamics of boundary work unfolds, in short: *Exclusion, Invitation* and *Disturbance. Exclusion* is the process whereby an actor is absented from a particular process of organizing. It can correspond to either the removal of actors outside a certain process of organizing, or their more absolute obliteration. Such boundary work can be spatial—actors occupy, and organize, in separate territories—or temporal—actors occupy, and organize, a space at different times. In contrast, *Invitation* is the process of strengthening relations between actors. This involves transforming their roles and potentials to act. Boundary work is herein concerned with new, or resurgent, forms of existence for all actors involved. The fate of one actor thus becomes tied to another (Callon 1986), engendering collective destinies. Lastly, *Disturbance* refers to the ongoing re-negotiation of roles and boundaries. This is why, for ANT proponents, no stabilization between actors can be guaranteed (Latour 1986). New actors may enter the organizing process, new interpretations may be given of events, tacit assumptions may turn out to be mistaken, and surprises may happen. Actors, such as animals, become 'troublemakers' that 'emerge in *surprising* fashion, lengthening the list of beings that must be taken into account' (Latour 2004, p. 79, emphasis added). Taken together, these three theoretical concepts make it possible to understand boundary work as it happens relationally *between* human and animal agencies.

Our two cases were developed for different reasons and at different time periods. When we began our study of the Scandinavian case company (DEF), we did not focus particularly on human/animal organizing as our overall aim was to do an explorative, open-ended study of relations between projects and firms. In line with ANT concepts of agency, we were interested in who or what made a difference in the project, how such differences were made and with what consequences (Latour 2005). This open approach allows surprises that, according to Latour (2013, p. 34), are unexpected elements arising from the empirical work that have to be added to the analysis. In an initial interview with a DEF project manager, we asked him what he thought were the biggest challenges in the project. To our surprise, he said that one of the most difficult things to handle had been the unexpected presence of about 500 moor frogs at a particular construction site and the effect this has had on the project's construction schedule. This led us to focus on the question of temporal organization and animals as one aspect of the larger case study. It became clear to us that animals could be seen as challenging organizational boundaries, and we decided to pursue this question of animals and organizational boundaries by exploring another case where we, in contrast to the Scandinavian case, knew in advance that the presence of animals was an issue. In that sense, the UK case, presented below, was a more planned exploration of this question with an explicit focus on animals and organizing in all interviews. The UK case was developed through an extant relationship with an infrastructure organization who provided us with access to data and people relating to a recently completed project where animals had become a concern. The two cases supplement each other by illustrating different aspects of how animals participate in organizing boundaries. Below, we give more detail about the two cases and the empirical material that informs our analysis.

The Scandinavian case concerned a project owned by a developer company DEF (a pseudonym) that was one of the big players in the Scandinavian construction industry in terms of revenue and profit at that time. The interviewed manager referred to a site in the suburbs of the capital city purchased by DEF with the aim of developing, building and selling about 80 hectares of new residential houses as well as renovating some existing buildings at the site. The building activities began in 2004, and the fieldwork on which the case study is based was conducted in 2006 while the construction activities were still going on. Our study was based on nine qualitative interviews (with DEF managers, employees and sub-contractors), a week's field study on the construction site where we followed the DEF project manager during his workday, including site inspections, participating in meetings, etc. It also included many informal conversations with the project manager and other people working at the project. In addition, we read numerous documents, including DEF annual reports, project plans, marketing material, consultant reports, documents from the local authorities and newspaper articles. These documents played an important role in our reconstruction of the frog case because they could reach further in time and space than any individual could recount.

The case located in the UK concerns an infrastructure project: the construction of a £45m value, 1200-mm high-pressure gas pipeline across 18.5 km from protected 'green belt' land into the outlying suburbs of a large conurbation. Construction took place between 2007 and 2010. The project client was a multinational, privately owned, regulated monopoly utility provider (named here 'Gasgen'). This project was unremarkable: it was not an environmental catastrophe and indeed received a sustainability award for waste management; however, it did involve the unusual construction of a large gas pipeline through a 'green finger' of uninhabited land into a large conurbation. To understand how animals, and their habitats, interacted with the organization of this project, between 2012 and 2014, we interviewed those most concerned with responding to their unpredictable presence within the project: two environmental managers within the client environmental division; the client's project environmental officer; the general contractor's environmental officer; the client's project manager; a sub-contacted environmental consultant; two local authority wildlife officers; and three people working for local wildlife charities involved in managing local wildlife reserves. These individuals were asked in semi-structured interviews to, *inter alia*, describe how wildlife was managed on the project, how wildlife shaped the project's development, and whether the project was typical of their experience of infrastructure projects. After the interviews, and post-construction, we walked through the site with several of these individuals to situate their experiences of the project and its ecologies. These field interviews helped us glimpse some of the complex animal agencies involved in the project more directly. Despite the historical nature of the case, many of the traces of the complex interactions between animals and humans on the project remained physically evident (e.g. new wildlife spaces and abandoned protective fencing). This data was thematically coded alongside documentation including: the environmental statement, environmental 'best practice' case studies prepared by the client and contractor, minutes of planning meetings within the local authority and environmental mitigation protocols. As with the DEF case, across our heterogeneous data set, thematic coding focussed upon tracing the different agencies that made a difference to the spacing and timing of the project.

For both cases, the empirical material was read and reread several times by all research team members in an open manner consistent with the agnostic and symmetrical approach of ANT where the agency of no actors is privileged in advance (e.g. Latour 2005). As we conducted our analysis of both cases, it became clear to us that the animals, and their habitats, could be seen as actors in the ANT sense of the term because they clearly made a difference to the organization of the projects in the question. On this basis, we read the material with the aim of tracing the different associations that these actors formed, how they were transformed and how they transformed each other, and how these agential relations affected the (human) organizing processes of the project. In that sense, our conceptual triad of Exclusion, Disturbance and Invitation were conceptualizations formed on the basis of a close reading of the material. Once formed, however, we reread the material from both cases with the aim of analysing the relations between organizing, boundary work and animals. If we had started with other actors and associations, e.g. those forged by environmentalists or been able to follow the animal themselves as events unfolded, our case stories may have revealed different stories. However, in acknowledging these access limitations, which derive from the access granted by the construction firms involved, it is important to stress the specific purpose of our paper is to understand how animals make a difference to *human* organizing, and especially management. By focusing upon how human, mostly management actors, registers these differences, our empirical evidence is commensurate with this specific task. As our cases will disclose, from the point of the view of our participants, in the Gasgen project boundary work was primarily oriented towards the spacing of animals to ensure the planned temporal organization of the project, whereas in the DEF case it was animal timings that become the principle concern to achieve the planned spatial organization of the project. Thus, while to a degree both cases demonstrate the significance of spatial *and* temporal organizing, they are mobilized here to separately illustrate and elaborate upon the significance of human/animal boundary work to spatial, and then temporal, organizing.

Organizing with Animal Spacings

Exclusion

Projects are typically bounded by their management of time, cost and quality. For animals, boundaries include temporal cycles of species, spatial boundaries of habitats and agential interactions between species and habitats that comprise ecosystems. In construction, spatial exclusions make it possible to separate these two ways of organizing, typically via physical fences, blockades and walls—mechanisms that prevent different agencies from disorganizing each other. But these processes of exclusion do not simply correspond to static boundaries between the 'inside' and 'outside' of organizations (Cooper 1986; Santos and Eisenhardt 2005); rather, they are relational outcomes of boundary work.

In the UK Gasgen project, the separation of project and animal organization was twofold; it involved: (i) the network stabilization of animals by project ecologists who could speak for them (Callon 1986) and (ii) ad hoc, or fluid, adaptations to animal agencies by the wider project team. Concerning stabilization, a series of practices were devised to exclude animals from project space including high wooden hoarding fences, the pruning of trees near to access tracks to prevent nesting birds from encroaching onto the site, the use of timber matting on access tracks to allow small animals to pass beneath, the placement of escape routes in pipeline trenches for stranded badgers, the availability of an ecologist to relocate animals found within the site boundaries, special vibrating tape hung between posts to scare away breeding birds from nesting within the site boundaries, temporary exclusion fencing around great crested newt ponds and the mowing of grass within the site boundaries in advance of construction to prevent the passage and nesting of reptiles (grass snakes, lizards, adders and slow-worms). These technologies relied upon the support of expert spokespeople (ecological professionals) through which the organizing of animals and their habitats could circulate and be consistently articulated to project professionals (Callon 1986). The strength of these circulatory articulations was themselves dependent upon alignments between a network of heterogeneous materials (Latour 1986), including environmental impact assessment audits, wildlife laws, legal penalties, company policies, agreed budgets for environmental protection, project plans, along with certified qualifications, field experiments, laboratory experiments and scientific papers. The circulation of animal agencies into project management was thus at least partly dependent upon a fragile network of practices of registering animal agencies elsewhere in time and space.

Other acts of exclusion were constituted by more emergent adaptations between humans and animals. For instance, in some areas, the lack of tall grass, and piles of bare earth, within the boundaries of the construction site attracted nesting skylarks despite bird scaring tape. This was met by ad hoc bird scaring by sub-contractors and park rangers, notably the instalment of noisy diesel generators in prominent locations and regular ground patrols. The success of this ad hoc measures required the project team to exchange agential qualities with animals in a quite fluid manner (Bear and Eden 2011), to effectively think like animals—how can they move, what do they like, what do they dislike, even what does *this* bird seem to like/dislike, and what can their bodies do?

The effect of these exclusions made it possible to allow the construction team to proceed *as if* animal agencies, spacings and timings were excluded from their organizing. But crucially, these spatial exclusions did not simply involve the reduction, or domestication, of animal agency. Rather, whether across a stabilized network or within more ad hoc adaptations, animal agencies were now registered and circulating *within* the organizing of project management—they made a difference—but they only made a difference in so far as to help further their own exclusion from project organizing. Human organization and animal organization were thus separated to ensure they do not disorganize each other. This separation was, of course, helped along by wildlife law, institutions and professionals that could stop construction had the animals been harmed. But the successful maintenance of these boundaries certainly relied on the enrolment, to a certain extent, of animal organizing inside the organization of project management, so that animal agencies became reliably positioned outside of construction activities. In other words, this paradoxical recursive entanglement and disentanglement of animal/human agencies prevented conflict and allowed project management to continue on its predetermined plan.

Invitation

The success in maintaining separation between project management and animal organizational had its limits. Sometimes animal agencies were invited into project management and changed the direction of the project. For example, one invitation happened during negotiations to cross the River Ox, a small river located next to a major urban highway. While most river crossings and protected riverine animals on the project had been excluded from project management by placing pipelines in bored tunnels under river habitats, on the River Ox the project team proposed to offset time delays and budget overruns elsewhere by employing open-cut excavation. This technique involved damming the river, rerouting its flow via temporary pipes and pumps, digging a trench, laying the pipeline and reinstating the river. Consent for this proposal was successfully granted by the statutory authority because Gasgen agreed to fund a specialist sub-contractor to undertake significant habitat enhancements on this section of the Ox. These enhancements included the removal of Himalayan balsam (an invasive plant that can overgrow river banks impeding native riverine plants and water vole habitats), the creation of a backwater waterhole as a habitat for fish, insects and amphibians, the instalment of flow deflectors, berms, gravel bars and pebble riffles to increase the oxygen content of the water to nurture fish and provide egg laying areas, and the re-grading of the river bank to enable easier access by riverine animals.

This example of boundary work is not simply one of exclusion, of separation, wherein animals make no, or little, difference to the predetermined organization of a project; this is because in this instance the boundaries of the construction project also changed. That is, the shape of the pipeline changed: its three-dimensional boundaries were disorganized—the depth of pipeline was reduced, saving project management time and money. Project management thus became dependent upon the simultaneous disorganization and reorganization of project parameters along with animals' habitats. The decision to open-cut the river was certainly detrimental, in the short term, to riverine animals—hence the original plan to recommend tunnel boring to avoid damage. However, this act helped secure the planned route of the project, as well as project deadlines, while also potentially at least producing spatially expanded habitats for riverine animals. In other words, this boundary work involved mediating between project management and animal organizing; project characteristics could only be changed if animal organizings were changed. Thus, human and animal agencies are entangled not simply to police their separate organizings; rather, the fate of their organizings becomes more profoundly tied to each other. What is more, this act of invitation cut both ways: animals were invited by humans to be tied to their organizing, just as humans were invited by animals to be tied to their organizing. Indeed, the very abundance of animal habitats in this area, and other areas along the pipeline route, invited the project into their habitats in the first place. This is because large diameter gas pipelines cannot, in the UK, be legally routed under buildings—the pipeline required their 'wild' space, just as animals may one day require the pipeline to protect themselves from urban sprawl. Such examples suggest how human and animal agencies become entangled in new ways, creating collective destinies.

Disturbance

Thus far, the Gasgen example has served to explain two different ways through which animal organizing makes a difference to (human) project management. Importantly, these two forms of organizing and their particular, indeed different, delimitations of agency, space and time were produced relationally through agential entanglements. We have proposed two concepts: Exclusion and Invitation to make sense of the different forms these entanglements might take. We would now like to propose a third: Disturbance. These acts offer a different agential starting point to those analysed thus far; namely, here we begin with animal organizing rather than project management, specifically a group of badgers. These badgers occupied a woodland sett discovered on the planned pipeline route. Natural England, the UK statutory body charged with enforcing wildlife legislation, stipulated that the badgers could not be harmed or removed from the sett as this would be illegal under the 1992 Badger Act; the project team were told to exclude the badgers from space by constructing an artificial sett outside the project boundaries and then using food to entice them to move into this sett. As with the discussion of exclusion above, the production of spatial boundaries by project management was contingent upon the extension of a stabilized network, through which the spokespeople for the badgers (Natural England) and their techniques of exclusion could travel into any place, at any time, essentially speaking for all badgers.

Despite repeated efforts to extend their ecological network, alongside more fluid ad hoc practices (experiments in food types and trails), the badgers' organizing, their spatial and temporal boundary work, remained unassimilable to project management. The badgers decided not to incorporate the new sett into their territory, and, owing to the presence of nearby housing, no other route was legally possible. Here, the badgers made a difference: they challenged their designated expert spokespeople (Natural England) by refusing the extension of Natural England's ecological network to speak for them, and their agency, into this woodland. But Natural England did not themselves lose interest in badgers, even if the badgers appeared disinterested in them. Instead, Natural England, instructed by the 1992 Badger Act, reiterated their refusal to allow the project team to move the badgers, that is, to disrupt the badgers' boundary work. Consequently, the project team had to disorganize the pipeline. This involved boring a tunnel beneath the original badger sett, an event which added around £500,000 to the project costs and delayed it for several weeks; however, these costs and the delay were to an extent mitigated by the (previously unplanned) opportunity to open-cut the River Ox later in the project.

In this example, the refusal by the badgers to be aligned with their spokespeople (Natural England) changed the spatial and temporal (and budgetary) limits of the project. In other words, they made a difference, they acted, but they did not *possess* agency. Rather, their power to refuse, their agency, was relationally constituted: it was contingent upon their interactions with their presumed spokespersons from Natural England, along with the 1992 Badger Act and technological objects like an artificial badger sett. Here, the transformative interplay between humans and animals, and other non-human objects, equips the badgers with 'their' agency to refuse (or not). But this refusal does not correspond to their exclusion from human organization or an invited conjoining with human organizing, rather their refusal disturbs, or disorganizes, human organizing in a less certain, asymmetrical, way. In the next section, we turn to our second construction case to illustrate how these three aspects of boundary work—Disturbance, Exclusion and Invitation—can also help to elucidate the different temporalities through which animals make a difference to project organizing. We start with another story of disturbance.

Organizing with Animal Timings

Disturbance

It came as a surprise, and at first an unwelcome one, to DEF when environmentalists made them aware that their construction site contained not only two, partly hidden ponds, but also about 500 moor frogs, a species protected by European Union Law (EC Directive 92/42). This discovery disorganized the project. Time, and in particular speed, was a very important aspect of DEF's business model (Tryggestad et al. 2013) as expressed in the following quote from an interview with DEF's CEO:

> You know, we're really fast. That's one of our parameters. We've always been 3-6 months faster than our competitors [...] That's why we hardly ever buy a piece of land where we have to struggle with the plans for 5, 6 and 7 years. Normally, we will buy something that is easy to approach—with a district plan process that is almost settled. We want to finish the project in the same market because the market fluctuates. So the speed of a project is a very important parameter for us.

It was important to DEF's managers to organize their project within a linear chronology with strict deadlines. The frogs threatened to disorganize the temporal boundaries set up around the project and its timings. This project organized time into a set of linear slices, a critical path of tightly interlinked project tasks, defined around the beginning and ending of this, and indeed most construction projects. The temporality of the frogs worked differently: it was orientated around the seasonality of their habitat and its interaction with their own agencies, especially their capacity to breed. For these actors, time was organized into temporal cycles: winter hibernation; breeding in early spring; spawning in ponds in late spring by females; tadpole metamorphosis into froglets over summer and early autumn; and then winter hibernation. The discovery of these actors started to disorganize the linear temporality of the project, as development around the waterhole was rescheduled: the project plan was delayed and now somehow had to take into account the cyclical time of the frogs; this dismayed the project managers in DEF. DEF's managers attempted to 'deal with' this disturbance to their organization of project time through practices of exclusion, but these exclusionary practices differed somewhat from those effected by Gasgen.

Exclusion

An apparent attempt to exclude the frogs from organizational space and time was reported in a local newspaper in the autumn 2004 when the police were contacted by an environmentalist. When the police arrived, they found several excavators digging near the ponds at the construction site and so brought work to a halt. After the incident, environmentalists claimed that DEF had deliberately tried to exterminate the frogs. The allegedly planned exclusion was quite unlike the sensitive negotiations with animal agencies practiced by Gasgen. Rather, it seemed to speak more to aspects of a managerialist control of 'Nature'. The cyclical time of the frogs was perceived as monstrous, an 'Other', to the linear time of project management. From the perspective of DEF's project management, the temporal monstrosity of the frogs corresponded not just to their disorganizing potential to threaten predetermined project deadlines, but also their lack of deadline. The frog's cyclical timing could not be represented in the linear, task end-date competition, temporality, of DEF's project management. By planning their extermination, the frogs were to be given a deadline, so they could become just another task on the critical path of DEF's project temporality.

But then, this project management became unstuck. DEF's management claimed to the environmentalists and police that that they had already received permission to remove the waterholes. In other words, they sought to justify the extermination of the frogs on the basis of an earlier time in which the frogs were said to not exist. Evidence presented by the environmentalists—the frogs' spokespeople—suggested a clear link between the waterholes and frogs. Given the controversy, DEF considered gaining legal permission to relocate the frogs to a nearby waterhole beyond the spatial boundaries of the construction site. DEF eventually realized that an application to the authorities to move the frogs, which would no doubt require protracted, and costly, negotiations with the frogs and their spokespeople, might jeopardize the project's linear time schedule even if DEF were to get approval.

Invitation

As a result of the delays, and the possibility of disorganization to the project schedule, DEF made a decision:

> So, we ended up turning things around, saying 'ok', instead of fighting, they [the frogs] should be allowed to live there. But we must build anyway, so we need to know something about how this kind of frog would like to live. Then we followed suit and recruited the country's leading moor frog experts as our advisors. (DEF project manager)

The frogs were now invited to make a difference to project management through their new spokespeople. This invitation disorganized and then reorganized the temporal boundaries of the project. Previously, time was predominantly perceived as linear and finite. The temporal boundaries were established with a stable beginning and an end, and the activity in the project was supposed to progress according to a pre-established time plan, manifest on plans and charts. However, after the frogs were invited in new temporal, concerns were introduced, changing the temporal organization of the project, as a DEF project manager explained in interview:

> There is a particular time schedule for handling the frogs. When it is their breeding season, they want to use the pond. Then a frog fence will be erected to keep them on the trail. There is a frog fence where they exit the waterhole so they don't escape.

Just as with the Gasgen project, the project was disorganized and reorganized once the frogs were invited into the space of project management: fences were erected to direct the behaviour of the frogs and make sure they were protected from the construction site machines and special frog corridors were built under the road so that frogs could wander relatively freely into and out of their habitat. But during the breeding season, between March and June, some barriers were opened and construction work near the waterhole was redirected and slowed to allow the frogs' unrestricted access to the water for breeding. As the above quote shows, DEF's time plan had to be supplemented by a different time plan that took the breeding cycle of the frogs into account. The different spatial, temporal and agential limits of project management and frog organization were not primarily just negotiated through reorganizing space as with our Gasgen case. In the DEF case, temporal boundaries came to the fore. The project was timed, as well as spaced, differently, because of its entanglements with animals, allowing different ways of organizing time not just to coexist but also to influence each other. Rather than being excluded by linear project time, as initially appeared to be likely, the cyclical timing of the frogs was allowed to make a difference to the timing of the project, resulting in a slower, but hopefully more viable, project. Equally the timing of the frogs to breed was now at least partly dependent upon their compliance on their timed arrival and departure for breeding within a predetermined plan within the linear charts of project management.

Concluding Discussion

Our two empirical cases illustrate the diversity and complexity of human and animal boundary work. The construction industry was mobilized here to empirically develop new insights into how animals and humans organize different spaces and times through their entangled relations. Given the pervasive role of the built environment in effecting separations between animals and humans (Philo and Wilbert 2000), we contend construction management is an important, and often overlooked (Sage 2013), field through which such geographies are relationally constituted. Just as with other practices such as agricultural domestication, (construction project) management organizing can thus shed light on the composite 'patchwork' of relational practices through which 'culture is divided from nature' (Lien and Law 2011, p. 82). Even within the most managerial organizings such as a construction project, animal agencies, spacings and timings make a difference to such separations. Our empirical examples thus challenge both managerialist (Santos and Eisenhardt 2005) and poststructuralist (Bloomfield and Vurdubakis 1999; Cooper 1986) notions that animals are passive, or monstrously 'Other', to the boundary making that delineates the orderliness of (human) organization. Rather, they are part of those ordering processes.

We proposed three concepts—*Invitation, Exclusion* and *Disturbance*—as heuristics to account for the relational boundary work that occurs between human/animal organizing. Each concept helped explain a distinctive process wherein human/animal agencies become constitutively entangled with another. 'Invitation' refers to the capacity of human animal agencies to interact and foster a collective, somewhat mutually dependant, fate for their different organizings. For example, the invitation of moor frogs, and their cyclical time, into the DEF project, served to allow the linear time of the project to continue, while also protecting the cyclical timing of the moor frogs. 'Disturbance' explains how unruly agencies may circulate and delimit the spacing or timing of another agency without warning. Temporally, this can bring a halt to project management (as with the DEF example), and spatially, 'Disturbance' can, quite asymmetrically, set new boundaries for one way of organizing. Finally, we also considered 'Exclusion', or separation, which is perhaps the most commonly acknowledged boundary practice between humans and animals. Spatially, as in the Gasgen case, 'Exclusion' concerns the boundary work that keeps human and animal ways of organizing, relatively, separate from each other, preventing mutual disorganization. Temporally, 'Exclusion' indicates boundary work where the timings of one way of organizing are forcibly excluded by the imposition of another. In the DEF case, this resulted in the frogs being almost subject to annihilation, as for them, timing was a matter of life and death, not simply cost efficiency.

Our argument highlights the processes of adjudication between agential, spatial and temporal boundary work. For humans to organize, that is, become concerned with working up agential, spatial and temporal limits (Latour 2013), this can involve complex entanglements with unruly animals, as well as technological agencies. We considered how human and animal organizing becomes entangled, so that a housing constructor times their working rhythms around the breeding patterns of frogs in order to build on schedule, or the survival of the habitats of a group of skylarks near a building site depends upon the extension of an ecological network that circulates the specification for vibrating bird scarring tape. Thus, animal organizing is not simply a negated 'Other' (Bloomfield and Vurdubakis 1999; Cooper 1986), beyond human organizing but rather a constituent part of that organizing. Finally, and perhaps most significantly, our analysis challenges the association of management (Cederström and Fleming 2012; Fleming and Spicer 2004; Quattrone and Hooper 2005; Parker 2014; Winiecki 2009), and especially project management (Law 2002b; Lundin and Söderholm 1994), with the effacement of unruly, or 'wild', agencies, spacings and timings. The entanglements of animal/human organizings we encountered suggest that management does not simply involve the effacement of 'wild' animal agencies, spacings and timings in a humanistic, totalitarian, quest over 'Nature' (Klikauer 2013; Newton 2009; Parker 2002). Management, and its mission to order space-time (Kornberger and Clegg 2004; Quattrone and Hooper 2005), even to control the boundaries of organization (Santos and Eisenhardt 2005), may instead sometimes be partly enacted *through*, rather than *against*, different, even disorganizing, certainly troublesome, but never monstrous or simply managed, animal agencies.

References

Anderson, K. (2014). Mind over matter? On decentering the human in Human Geography. *Cultural Geographies, 21*(1), 3–18.

Bear, C., & Eden, S. (2008). Making space for fish: The regional, network and fluid spaces of fisheries certification. *Social and Cultural Geography, 9*(5), 487–504.

Bear, C., & Eden, S. (2011). Thinking like a fish? Engaging with nonhuman difference through recreational angling. *Environment and Planning D: Society and Space, 29*(2), 336–352.

Bloomfield, P., & Vurdubakis, T. (1999). The outer limits: Monsters, actor-networks and the writing of displacement. *Organization, 6*(4), 625–647.

Callon, M. (1986). Some elements of a sociology of translation: Domestication of the scallops and fishermen of St. Brieuc Bay. In J. Law (Ed.), *Power, action and belief: A new sociology of knowledge?* (pp. 196–233). London: Routledge.

Cederström, C., & Fleming, P. (2012). *Dead man working*. Winchester: Zero Books.

Clegg, S., Kornberger, M., & Rhodes, C. (2005). Learning/becoming/organizing. *Organization, 12*(2), 147–167.

Clegg, S., Kornberger, M., Carter, C., & Rhodes, C. (2006). For management? *Management Learning, 37*(1), 7–27.

Collins, H., & Yearly, S. (1992). Epistemological chicken. In A. Pickering (Ed.), *Science as practice and culture* (pp. 301–326). Chicago: University of Chicago Press.

Cooper, R. (1986). Organization/disorganization. *Social Science Information, 25*(2), 299–355.

Deleuze, G., & Guattari, F. (1987). *A thousand plateaus: Capitalism and schizophrenia*. London: Continuum.

Fleming, P., & Spicer, A. (2004). 'You can checkout anytime, but you can never leave': Spatial boundaries in a high commitment organizations. *Human Relations, 57*(1), 75–94.

Ginn, F. (2014). Sticky lives: Slugs, detachment and more-than-human ethics in the garden. *Transactions of the Institute of British Geographers, 39*(4), 532–544.

Grazian, D. (2012). Where the wild things aren't: Exhibiting nature in American zoos. *The Sociological Quarterly, 53*(4), 546–565.

Greenhough, B., & Roe, E. (2011). Ethics, space, and somatic sensibilities: Comparing relationships between scientific researchers and their human and animal experimental subjects. *Environment and Planning D: Society and Space, 29*(1), 47–66.

Hassard, J., Keleman, M., & Cox, J. (2008). *Disorganization theory: Explorations in alternative organizational analysis*. London: Routledge.

Holloway, L. (2007). Subjecting cows to robots: Farming technologies and the making of animal subjects. *Environment and Planning D: Society and Space, 25*(6), 1041–1060.

Jones, G., McLean, C., & Quattrone, P. (2004). Spacing and timing. *Organization, 11*(6), 723–741.

Keul, A. (2013). Embodied encounters between humans and gators. *Social and Cultural Geography, 14*(8), 930–953.

Klikauer, T. (2013). *Managerialism: A critique of an ideology*. London: Palgrave Macmillan.

Knox, H., O'Doherty, D., Vurdubakis, T., & Westrup, C. (2015). Something happened: Spectres of organization/disorganization at the airport. *Human Relations, 68*(6), 1001–1020.

Kornberger, M., & Clegg, S. (2004). Bringing space back in: Organizing the generative building. *Organization Studies, 25*(7), 1095–1114.

Latour, B. (1986). *Science in action: How to follow scientists and engineers through society*. Cambridge, MA: Harvard University Press.

Latour, B. (2004). *The politics of nature*. Cambridge, MA: Harvard University Press.

Latour, B. (2005). *Reassembling the social*. Oxford: Oxford University Press.

Latour, B. (2013). *An inquiry into modes of existence*. Cambridge, MA: Harvard University Press.

Law, J. (2002a). Objects and spaces. *Theory, Culture & Society, 19*(5/6), 91–105.

Law, J. (2002b). *Aircraft stories: Decentring the object in technoscience*. Durham, NC: Duke University Press.

Law, J., & Lien, E. (2013). Slippery: Field notes in empirical ontology. *Social Studies of Science, 43*(3), 363–378.

Law, J., & Mol, A. (2008). The actor-enacted: Cumbrian sheep in 2001. In C. Knappett & L. Malafouris (Eds.), *Material agency* (pp. 57–77). New York: Springer Science and Business Media.

Lien, M., & Law, J. (2011). 'Emergent aliens': On salmon, nature, and their enactment. *Ethnos: Journal of Anthropology, 76*(1), 65–87.

Lorimer, J. (2008). Living roofs and brownfield wildlife: Towards a fluid biogeography of UK nature conservation. *Environment and Planning A, 40*(9), 2042–2060.

Lundin, R., & Söderholm, A. (1994). A theory of temporary organization. *Scandinavian Journal of Management, 11*(4), 437–455.

Miele, M. (2011). The taste of happiness: Free-range chicken. *Environment and Planning A, 43*(9), 2076–2090.

Milne, M., & Gray, R. (2013). W(h)ither ecology? The triple bottom line, the global reporting initiative, and corporate sustainability reporting. *Journal of Business Ethics, 118*(1), 13–29.

Mol, A., & Law, J. (2005). Boundary variations: An introduction. *Environment and Planning D: Society and Space, 23*(5), 637–642.

Murdoch, J. (1998). The spaces of actor-network theory. *Geoforum, 29*(4), 357–374.

Newton, T. (2009). Organizations and the natural environment. In M. Alvesson, T. Bridgman, & H. Wilmott (Eds.), *The Oxford handbook of critical management studies* (pp. 125–143). Oxford: Oxford University Press.

November, V., Camacho-Hübner, E., & Latour, B. (2010). Entering a risky territory: Space in the age of digital navigation. *Environment and Planning D: Society and Space, 28*(8), 581–599.

Parker, M. (2002). *Against management*. Cambridge: Polity Press.

Parker, M. (2014). University Ltd.: Changing a business school. *Organization, 21*(2), 281–292.

Philo, C., & Wilbert, C. (Eds.). (2000). Animal space, beastly places: An introduction. In *Animal space, beastly places: New geographies of human–animal relations* (pp. 1–34). London: Routledge.

Quattrone, P., & Hopper, T. (2005). A 'time' space odyssey: Management control systems in two multinational organisations. *Accounting, Organizations and Society, 30*(7–8), 735–764.

Riach, K., & Kelly, S. (2015). The need for fresh blood: Understanding organizational age equality through a vampiric lens. *Organization, 22*(3), 287–305.

Sage, D. (2013). 'Danger building site—Keep out!?': A critical agenda for geographical engagement with contemporary construction industries. *Social and Cultural Geography, 14*(2), 168–191.

Sage, D., Dainty, A., Tryggestad, K., Justesen, L., & Mouritsen, J. (2014). Building *with* wildlife: Project geographies and cosmopolitics in infrastructure construction. *Construction Management and Economics, 32*(7–8), 773–786.

Santos, F. M., & Eisenhardt, K. M. (2005). Organizational boundaries and theories of organization. *Organization Science, 16*(5), 491–508.

Schreyögg, G., & Sydow, J. (2010). Crossroads—Organizing for fluidity? Dilemmas of new organizational forms. *Organization Science, 21*(6), 1251–1262.

Tajima, K. (2003). New estimates of the demand for urban green space: Implications for valuing the environmental benefits of Boston's big dig project. *Journal of Urban Affairs, 25*(5), 641–655.

Thanem, T. (2006). Living on the edge: Towards a monstrous organization theory. *Organization, 13*(2), 163–193.

Tryggestad, K., Justesen, L., & Mouritsen, J. (2013). Project temporalities: How frogs can become stakeholders. *International Journal of Managing Projects in Business, 6*(1), 69–87.

Winiecki, D. (2009). The call centre and its many players. *Organization, 16*(5), 705–731.

Whittle, A., & Spicer, A. (2008). Is actor-network theory critique? *Organization Studies, 29*(4), 611–629.

Index

© The Editor(s) (if applicable) and The Author(s) 2018
D.J. Sage and C. Vitry (eds.), *Societies under Construction*,
https://doi.org/10.1007/978-3-319-73996-0

Printed by Printforce, the Netherlands